Imposing Wilderness

California Studies in Critical Human Geography

Imposing Wilderness

Struggles over Livelihood and Nature Preservation in Africa

Roderick P. Neumann

UNIVERSITY OF CALIFORNIA PRESS

Berkeley / Los Angeles / London

University of California Press
Berkeley and Los Angeles, California

University of California Press, Ltd.
London, England

First paperback printing 2002

Library of Congress Cataloging-in-Publication Data

Neumann, Roderick P., 1954–
    Imposing wilderness : struggles over livelihood and nature
preservation in Africa / Roderick P. Neumann.
        p. cm.
    Includes bibliographical references (p.   ) and index.
    ISBN 978-0-520-23468-0 (pbk. : alk. paper)
        1. Arusha National Park (Tanzania).   2. National parks
and reserves—Social aspects—Tanzania—Arusha National Park.
3. Nature conservation—Social aspects—Tanzania—Arusha National
Park.   I. Title.
SB484.T3N48   1998
333.78'09678—dc21   98—12746
                                                                CIP

Printed in the United States of America

08   07   06   05   04   03   02

9   8   7   6   5   4   3   2   1

The paper used in this publication meets the minimum requirements
of ANSI/NISO Z39.48-1992 (R 1997) (*Permanence of Paper*). ♾

To my mother, Mary Neumann, and in memory of my sister, Christina Neumann

Contents

Illustrations

Figures

Maps

Tables

Acknowledgments

This book would have been impossible without the cooperation of the Meru families living near Arusha National Park. They not only tolerated and responded to what must have sometimes seemed the most inane questions, they opened their homes to me and treated me as a welcome guest. I am particularly indebted to Mathias Kaaya and the families of Sylvanus Kaaya, Felix Kaaya, and Serakieli Urio for their gracious hospitality and help in facilitating my research. Special thanks and credit go to Alawi Msuya, whose formal title was research assistant but who often served as my guide in an unfamiliar place and, ultimately, as an important source of information and insight.

I am also indebted to the staffs at several Tanzanian institutions for their assistance and hospitality. Numerous individuals at the Tanzania National Parks Headquarters, Arusha National Park, and the Serengeti Wildlife Research Institute patiently responded to my inquiries and often directed me to information sources and lines of investigation that I might otherwise have overlooked. The cooperation of Tanzania National Parks in allowing me access to their files and records was especially important to the study. Also deserving of special note are the staff members of the Tanzania National Archives.

Many friends and scholars have contributed ideas, encouragement, and inspiration during the various stages of this study, from the first research proposal draft to the final proofreading of the manuscript. At the risk of overlooking one or two I would like to thank Denis Cosgrove, Louise Fortmann, Eric Hirsch, Bernard Nietshmann, Brian

Page, Nancy Peluso, Rick Schroeder, James Scott, Jody Solow, Richard Walker, and the anonymous reviewers at the University of California Press. Special thanks to Gail Hollander for sharing her boundless enthusiasm for scholarly inquiry and all things Floridian. Her insights and encouragement (not to mention the book's title) were offered at crucial points in the manuscript revision process. I am also thankful for the suggestions, support, and criticisms received from the participants in the Program in Agrarian Studies, Yale University, where I was Postdoctoral Fellow in 1994–95. Tina Espinosa developed the maps contained within. Above all, I am grateful to Michael Watts, to whom I owe intellectual and professional debts of incalculable proportions.

A number of institutions have provided funding for various stages of this study. The main period of fieldwork in Tanzania was supported by the Joint Committee on African Studies of the Social Science Research Council and the American Council of Learned Societies, with funds provided by the Rockefeller Foundation and the William and Flora Hewlett Foundation, the Fulbright-Hays Doctoral Dissertation Research Abroad Program, and a National Science Foundation, Doctoral Dissertation Improvement Grant. Research at collections in England was funded by the National Endowment for the Humanities and the College of Arts and Sciences, Florida International University. A Postdoctoral Fellowship at the Program in Agrarian Studies, Yale University in 1994–95 allowed me to write or rewrite a significant portion of the book.

Finally, I wish to thank the staff at the University of California Press, particularly Mimi Kusch and Jean McAneny, for overseeing the editing of this book.

Introduction

Displeasing Prospects for People and Wildlife

"This is the way Africa *should* look." As my traveling companion made his declaration, he stood before an expansive East African landscape wearing a slightly ironic smile. We had driven together into the heart of one of Tanzania's renowned national parks, leaving behind the Maasai curios shops, the herds of scrawny cattle, and the desiccated maize fields. That *other* landscape—including the woman with the broken smile who, reeking of *pombe* (local beer) and shouting "*Zawadi!*" (gift), had thrust her hand through the vehicle window— was waiting just outside the park gate. Inside, the prospect was more pleasing. From our perspective above the banks of a placid, greenish river, the savanna stretched out before us in all directions into the haze of distant mountains. There were no obvious signs of human life or activity in the landscape, save the dusty ruts we had just followed through the park. This is the way Africa should look. The statement, from a British expatriate ecologist serving as an official of an influential international conservation organization, raises a number of questions central to this book. Among them, who decides what Africa "should" look like? Where and how have ideas of the pleasing African landscape been constructed? What does this landscape vision mean for African peasants and pastoralists living and laboring there? To what degree and in what ways do they resist this vision?

I pursue these questions by focusing on a small, forested national park, Arusha, in the northern highlands of Tanzania, and the Meru peasants who live, cultivate, and herd livestock on its boundaries. Encompassing 137 square kilometers of the highest slopes of Mount Meru, Arusha National Park has long been prized by European conservation advocates for its scenic value. Described by an early park scientist as "one of the most beautiful areas in Tanganyika" (Vesey-FitzGerald 1967, 11) and "a glimpse of Eden" by author-poet Evelyn Ames (1967), Mount Meru embodies the ideal of the picturesque natural landscape (see figure 1). Tanzania National Parks (TANAPA) has incorporated this ideal into its management policy, declaring that "[t]hese landscapes must remain unspoiled, as benchmarks to what once was" (TANAPA 1994, 1).

The idea of Mount Meru as picturesque landscape is, however, overlain by and implicated in an historical struggle over land and resources. Much of the Meru's territorial land claim was alienated by German and British colonial governments for settler estates or reserved for forest and wildlife conservation. Conflict over land and resources defined politics within Meru society and between the Meru peasantry and the state throughout the colonial era. Eventually, Meru peasant resistance to a colonial land alienation and resettlement scheme in the 1950s helped spark Tanzania's incipient nationalist movement. The struggles over land on Mount Meru were emblematic of local resistance to colonialism and broader efforts of nationalist parties to liberate their countries from colonial rule. Rather than representing a picturesque ideal, the landscape of Mount Meru locally embodies a decades-long struggle to defend and regain ancestral land claims.

Arusha National Park (and the forest and game reserves that preceded it) both contributes to and suffers from this history of land struggles. Since its designation in 1960, hostilities between park authorities and local communities have peaked and ebbed but patterns have remained consistent. From the perspective of park officials and wildlife conservationists, the conflict is defined by livestock trespass, illegal hunting, wood theft, and the consequent ecological costs such as species extirpation. For local Meru communities the conflict revolves around reduced access to ancestral lands, restrictions on customary resource uses, and the predation of wildlife on cultivated lands. There have been recurring confrontations over boundary locations and demarcations, access to local livelihood resources, and the enforcement of park and conservation laws. It is a conflict, I will argue, with deep

Figure 1. A portion of Arusha National Park rising above Nasula Kitongoji. (R. P. Neumann.)

historical roots in European colonialism and European ideals of the scenic African landscape.

It is also a conflict whose basic outline is common throughout Tanzania and indeed much of sub-Saharan Africa. While Arusha National Park has a unique history of human occupancy and use, administrative control, and ecological change, it encapsulates many of the contemporary conflicts surrounding the establishment and administration of protected areas across the continent. Seen from the perspective of international conservation organizations such as the World Wide Fund for Nature (WWF; the organization was formerly known as the World Wildlife Fund) and the World Conservation Union (IUCN), the conflict over protected area is defined by illegal hunting, grazing trespass, and boundary encroachment. Several game reserves in Tanzania, for example, are listed by the IUCN Commission on National Parks and Protected Areas as "threatened" by "intense poaching of certain species [and] . . . two sites are also suffering from takeover by pastoralists" (IUCN 1991, 299). The parks and reserves are viewed as remnants of a vacant African landscape where Nature was "unspoiled" until recent times. As a WWF African regional director recently wrote, the larger problem is the "transformation of formerly natural ecosystems for human purposes" (Lamprey 1992, 10). Conservationists have historically viewed the establishment and defense of national parks as the principal means of halting this process in Africa. Since 1960, the amount of land under strict protection for conservation has roughly doubled. Several countries, including Botswana, Namibia, Senegal, Tanzania, and Togo, have at least 11 percent of their territories in national parks and reserves.[1]

The great geographic expansion of protected areas on a continent whose population is primarily rural and agrarian has produced a conflictual relationship between people and parks. With the exception of Senegal, the populations of the countries listed above are 70 to 80 percent rural and mostly dependent upon agriculture and herding for their livelihoods. Establishing protected areas in Africa has meant the loss of access to land and natural resources by peasant communities who rely directly upon them for their survival. Protected areas in Tanzania, for instance, encompass a far greater proportion of land than in most industrialized countries. More than 25 percent of Tanzania's land is under some form of central government protection where cultivation and settlement are prohibited, including almost 14 percent in national parks and game reserves (IUCN 1991) (see map 1). Yet Tanzania is generally ranked among the ten poorest countries in the world, where approximately 80 percent of agricultural production is conducted by peasant households on plots averaging less than 2.2 hectares (DANIDA 1989, 42). In such national socioeconomic contexts across Africa, the loss of local land and resource access in the name of conservation has fueled rural conflict.

Rather than being "unspoiled benchmarks," most of the continent's protected areas have been created out of lands with long histories of occupancy and use. Forced relocations and curtailment of resource access continues apace around the region. In 1988 in Tanzania, five thousand pastoralists were forced out of the Umba-Mkomazi Game Reserve complex after refusing to obey a government eviction order. In Madagascar, cultivation was outlawed in the late 1980s when peasant agricultural lands were incorporated into Mananara Biosphere Reserve and grazing, fuelwood collection, and other activities were prohibited (Ghimire 1994). In Uganda, peasants were recently evicted en masse to create a wildlife corridor between the Kibale Forest Reserve and the Queen Elizabeth National Park (Colchester 1994). In essence, the

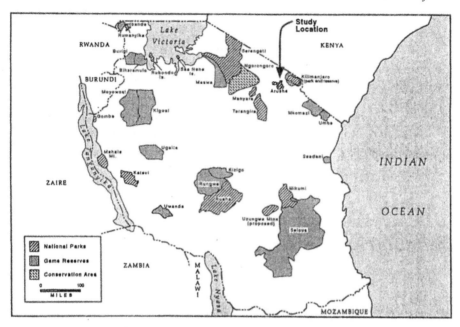

Map 1. Protected areas in contemporary Tanzania. (WWF 1990.)

establishment of national parks and associated protected areas criminal-
ized many customary land and natural resources uses for communities
across Africa. As a result, most protected areas have become arenas for
struggles over resources between state conservation agencies and the
local peasants and pastoralists.

The conflictual rhetoric is heating up and the level of antagonisms
shows signs of escalating. Several developments in Africa—particularly
the rise, in different locales, of both democratization and state repression, the weakening of central state authority, and the socioeconomic
strains of structural adjustment—have intensified tensions. Peasants,
grassroots activists, politicians, and social scientists increasingly question the relevance of national parks to the lives of rural Africans (see
e.g., KIPOC 1992; Caruthers 1994; Lowry and Donahue 1994; Ole
Ntimana 1994). In Tanzania, local political activists recently declared
national park policies to be a violation of human rights because of
their effects on rural livelihoods (Neumann 1995a). In Montagne
d'Ambre National Park in Madagascar, restrictions on land and
resource use in the late 1980s led to violent protests and arrests in
affected communities (Ghimire 1994). When the central government

of Togo began to weaken in 1991, rural residents, angered by an historical pattern of repression and dislocation, chased away national park officials and proceeded to reoccupy Keran National Park (Lowry and Donahue 1994).

One of the African governments' responses to these threats has been to step up the level of state violence in the name of conservation. A report from the Botswana Christian Council claimed that "Bushmen" suspected of hunting on their former lands were sadistically tortured by wildlife and parks officers (Kelso 1993).[2] In Togo, there have been accusations of the park service using helicopters to shoot poachers in Keran National Park (Lowry and Donahue 1994). French soldiers admitted killing wounded poachers brought down by their guns in the Central African Republic (Colchester 1994). As tensions rise, a general trend toward militarizing protected areas is evident. Tanzania's National Park agency has created a "paramilitary" unit "governed by the paramilitary disciplinary code of conduct" (TANAPA 1994, 63) to defend its protected areas against local communities. In some cases, such as Kenya (Peluso 1993) and southern Africa (Ellis 1994), the militarization of conservation has overlapped with state repression of minority ethnic groups and liberation movements.

Antagonisms between African states and rural communities over conservation policies are high, in part, because there is a great deal at stake economically and ecologically. Nature tourism is one of the top foreign exchange earners for several sub-Saharan countries, including Botswana, Zimbabwe, and Kenya. In Tanzania, the system of national parks and protected areas is the key attraction for both foreign investors and international tourists. Tourism was the country's fastest-growing industry in the first half of the 1990s, fueled by a large influx of foreign capital (Neumann 1995a). The government hopes to more than double foreign exchange earnings, from $205 million in 1995 to $570 million in 2005 (EIU 1996). National politicians and international financial advisers view revenues from nature tourism as vital to an economy burdened by $8 billion in external debt. The continued interest of investors is, to an important degree, dependent on the suppression of conflicting claims to land and resources at Tanzania's ecotourism destinations.

In ecological terms, conservationists see national parks as the final sanctuaries for threatened wildlife populations and their habitats. Most visibly, the continent-wide populations of African elephant (*Loxodonta africana*) and black rhino (*Diceros bicornis*) dropped precipitously in the 1980s (see e.g., Ricciuti 1993). Tanzania is of particular concern to international conservationists, not only because of the status of its elephant and rhino, but for its ecological richness in general. Superlatives dominate their accounts of its flora and fauna. A recent World Wide Fund for Nature document calls Tanzania "one of the most important countries in the world for conservation" (WWF 1990, 3). Its forests are "of great biological importance," its major parks have "outstanding universal value," and its coral reefs are "among the richest in the world" (7–10). Despite the state having set aside extensive areas for protection, conservationists in previous decades witnessed a steep decline in some of Tanzania's wildlife populations. In the mid-1980s, its elephant population dropped by more than 50 percent. Illegal hunting decimated 98 percent of the black rhino population so that only a few hundred remain. In cases where wildlife species are in danger of extirpation or even extinction, some conservationists believe that the only recourse is to essentially declare war on the groups and individuals deemed responsible (Peluso 1993; Bonner 1993).

Contextualizing Protected Area Conflicts

While state violence is pervasive and even escalating in particular locales, there is a growing movement within international conservation organizations and national agencies to implement a "new approach" to conservation in Africa (see Neumann 1997a). Local resistance to the loss of land and resource access has pushed conservationists and state officials to reassess coercive park and wildlife protection policies. During the first two decades of independence, peasants and pastoralists relentlessly confronted the newly trained African conservation officials with challenges to park and wildlife laws. Additionally, writers from a variety of social science disciplines began in the 1980s to sharply criticize the implementation of national park policies for their disregard of local property claims and human rights (see e.g., Marks 1984; Arhem 1985; Collett 1987; Turton 1987). Stung by these criticisms and faced with a hostile and increasingly militant rural populace, conservationist literature began to emphasize the need to reconcile wildlife conservation with the needs of local people (see, e.g., Dasmann 1984; McNeely and Miller 1984; Miller 1984; McNeely and Pitt 1985; Kiss 1990). It has since become de rigueur to emphasize "local participation" and "community development" as key to the future of nature protection (see, e.g., Wells and Brandon 1992; Omo-Fadaka 1992; Cleaver 1993; Baskin 1994).

The new emphasis on local participation and benefit sharing in conservation is an important shift in thinking, and it may help to alleviate the conflicts that wrack protected areas in Africa. As many of the new participatory projects take shape, however, there are signs that some programs paradoxically facilitate greater state surveillance and control of local land and resource use (Lance 1995; Hill 1996; Neumann 1997a). The source of the paradox can be found in the prevailing explanation for the problems that the new programs are meant to alleviate. There is a standard twofold explanation for conflicts that emerges repeatedly in conservationist literature: increasing human populations and a lack of local understanding of the importance of wildlife conservation. The WWF regional director in Nairobi recently wrote that environmental degradation resulted from "expanding human populations and their increasing demand for land" (Lamprey 1992, 9). A recent IUCN document defines the problem in Tanzania explicitly: "Within protected areas specifically, there are conflicts between the needs of the parks and of the local people as populations increase" (IUCN 1991, 301). A UNESCO scientist combines this explanation with a concern about the lack of understanding of conservation: "I am convinced that the majority of people who are being added to the world's population today do not have an understanding or sympathy for conservation" (Lusigi 1992, 79). Conservationists working in Tanzania have long emphasized the need to educate local villagers on the value of conservation and national parks. A TANAPA progress report for 1972 to 1975 declares that one of the "main activities in management is conservation education . . . on the proper use and value of wildlife."[3] At the country's most well-known national park, Serengeti, a new regional management plan includes "education in natural history and wildlife management among the local villages . . . so that they can appreciate the wildlife spectacle" (Mbano et al. 1995, 609).

Though these explanations of protected area conflicts have merit, they greatly limit the type of analyses possible. The IUCN survey cited above (IUCN 1991), for example, makes no reference at all to the political, socioeconomic, or historical context of protected areas in Tanzania. National parks are treated as threatened by land-hungry populations rather than as being historically complicit in the creation of conditions of land and resource scarcity. My purpose herein

is to analyze the conflicts between nature protection and rural livelihoods in Africa within their historical and sociopolitical context. In the
chapters that follow, I conceptualize national parks not simply as threatened by social, political, and economic forces beyond their control, but
as active sociopolitical forces in their own right. Parks and protected
areas are historically implicated in the conditions of poverty and underdevelopment that surround them. Through historical contextualization, I challenge standard interpretations of the conflicts surrounding
parks as being driven by population growth and ignorance of conservation values. In a detailed analysis of Arusha National Park, I demonstrate the effects of nature protection on rural livelihoods and document the historical continuity in local responses to state preservation
policies.

Drawing from political ecology, as well as theories of peasant society
and social constructions of nature, the book reveals the underlying
sociopolitical motivations in many of the violations of resource laws. My
analytical interests lie in the symbolic importance of landscapes, the
political struggles over landscape meanings among different social
groups, and the confluence of struggles over meaning and struggles
over land and resource access. The European appropriation of the
African landscape for aesthetic consumption is inseparable from the
appropriation of African land for material production. In the analysis
that follows, I investigate the origins of conflicts between Arusha
National Park and surrounding communities by historically contextualizing the development of the national park ideal in Tanzania.

By "national park ideal," I am referring principally to the notion that
"nature" can be "preserved" from the effects of human agency by legislatively creating a bounded space for nature controlled by a centralized
bureaucratic authority. This model was initially implemented in the
nineteenth-century United States at Yellowstone (present-day Wyom-
ing and Montana) and has subsequently been disseminated worldwide.
In Africa, European colonial authorities were responsible for establishing the Yellowstone model of national parks, which in many ways
helped to legitimate and reinforce imperial rule (D. Moore 1993;
Ranger 1989; Caruthers 1989; 1994; Neumann 1996). I also use the
term *national park ideal* to refer to a particular conceptualization
of nature that has been prevalent in the Yellowstone model since its
inception. It is a conceptualization of nature that is largely visual, that
treats nature as "scenery" (see Nash 1982, 112–15) upon which aesthetic judgments can be laid. Fundamental to the structure of the

national park ideal is an Anglo-American nature aesthetic whose formulation we can trace historically.

My aim is neither to essentialize national parks as simply the manifestation of a set of aesthetic values nor to confuse the ideal with the complex and dynamic motivations that have powered the nature preservation movement globally. Rather it is to explicitly acknowledge that we—meaning "educated Westerners"—recognize certain landscapes as natural in part because we have been trained to expect a particular vision through centuries of painting, poetry, literature, and landscape design. In other words, my aim is to acknowledge that the idea that nature can be preserved in parks is culturally and socially produced in a way that can be traced to documented historical processes in specific locales. While the analysis that follows is thus dependent on a close examination of the Anglo-American nature aesthetic, it also investigates the potential for dissenting visions. I am thus concerned with how peasants and pastoralists have challenged the scenic vision of Africa and its implementation in the form of national parks. This is not to say that I will present a point-by-point comparison of a "Meru" cultural vision of nature vis-à-vis an "Anglo-American" one. Instead I investigate and explain how the implementation of the national park ideal affected local patterns of livelihood among Meru peasants and what actions they have taken to accommodate, resist, and mitigate the negative effects. The landscape of Arusha National Park, then, provides the focus to critically examine the ideal as applied in Africa and to uncover local patterns of resistance, both to the criminalization of former land use practices and to the loss of customary rights of access.

Overview

Much of the book is based on my two periods of doctoral dissertation fieldwork in Tanzania in 1988 and 1989–90. Additional archival research was conducted in England in 1993. I employed a variety of methods, including household surveys, archival research, interviews, and on-site observation. In general I sought to contextualize, historically and politically, the actions and speech of Meru peasants living near the boundary of the park. By approaching the analysis in this way, I have been able to examine the sociopolitical motivations behind common violations of park and wildlife laws. Because of the consequences for villagers should they be found by government authorities to be engaged in illegal resource use, I have taken precautions in conducting and writing about the research. Respondents to the survey were guaranteed anonymity, and I have created pseudonyms for interviewees and otherwise masked people's identities.

The larger part of my interviews and observations were taken in two *vitongoji* (subvillages, plural form) on the park boundary (see map 2). I chose these particular communities because they are directly adjacent to the park's boundary. Other villages are further removed because much of the park is surrounded by a government forest reserve. I chose one *kitongoji* (an administrative subdivision of the village), Nasula, on the northern boundary and one, Ngongongare, on the southern boundary in order to achieve some socioeconomic and ecological diversity that might allow me to make comparisons. The areas are indeed very different but are also similar in ways that I did not realize when I chose them. Significantly, both are located on lands originally alienated from Meru territorial claims by colonial governments in order to create European estates.

The first chapter lays out a theoretical framework for the study which combines the literatures on landscape and social constructions of nature with theories of peasant resistance. It begins by demonstrating that the social history of landscape and its role in social constructions of nature are critical to understanding the development of the national park concept. I argue that national parks are historically and culturally contingent representations of a particular nature aesthetic. Parks are landscapes of consumption, upon which are projected ideas of culture and nature and of where (literally) to draw the boundary between them. Following this argument, I demonstrate the important historical and contemporary role that conservation and national park laws play in reinforcing and legitimating state control over land and resources in rural Africa. I close by arguing for a geographically and historically specified concept of landed moral economy as a key analytical tool for understanding the character of local responses to the loss of customary property rights to the state in the name of nature preservation.

The next chapter begins with a brief history of pre-European settlement and land use to initiate an analysis of how European colonial occupation challenged existing symbolic and material uses of Mount Meru. It traces the historical transformation of the local political economy through land alienations for European commercial estates and the parallel process of Meru peasantization. It demonstrates how struggles over land and natural resource access rights have defined politics on Mount Meru for close to a century. The history of the land crisis pro-

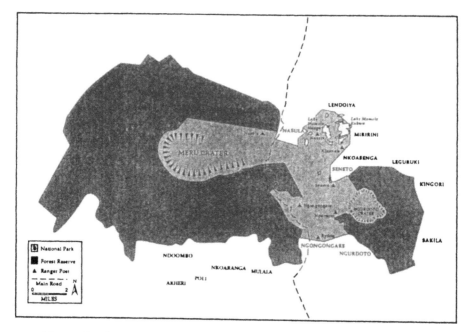

Map 2. Arusha National Park and surrounding communities. (Snelson 1987
with additional information compiled by R. P. Neumann.)

vides the context for understanding contemporary patterns of uneven
development on the mountain. Using interview and survey data, the
chapter closes with a detailed examination of the settlement histories,
land tenure systems, and household production strategies of three representative villages, two of which lie on the boundary of the Arusha
National Park.

This is followed by an examination of the historical process whereby
the state extended control over access to and benefits from forest and
wildlife resources in Tanzania. Chapter 3 thus identifies a trend of geographically expansive and increasingly restrictive policies governing
resource use from the German colonial period through to independence.
Through mainly archival sources, it identifies a pattern of community
resistance on Mount Meru that remains remarkably consistent over the
course of roughly four decades of British colonial rule. The chapter examines the effects of state policies from the perspective of Meru peasants
who claim that the Meru Forest and Game reserves and national park
usurped customary land and resource rights. An explanation of this historical pattern is critical to understanding contemporary conflicts
between the park authorities and surrounding Meru communities.

Chapter 4 opens with a social-historical analysis of the world's first international wildlife conservation organization, the London-based Society for the Preservation of the Fauna of the Empire (SPFE). The interventions of this group were critical in shaping the program to establish a system of national parks in Tanzania from 1930 to the late 1950s. I argue that the SPFE's aristocratic leadership combined their own privileged experience of English game laws and landscaped parks with ideas of an African Eden to structure their model for national parks. The chapter traces the political struggles among various segments of the colonial administration, the SPFE, and affected Africans through the history of Tanzania's first national park, Serengeti. In doing so, it places national parks at the center of larger struggles for land and resistance to colonial occupation. The latter part of the chapter examines the transition to postcolonial nature protection policies, including the establishment of the first postcolonial park, Arusha.

The principal purpose of chapter 5 is to provide a detailed description of contemporary antagonisms between the national park administration and surrounding communities. In it, I identify "patterns of predation" from the perspectives of both the park administration and local villagers. That is, spatial, temporal, and social patterns of common park violations such as grazing trespass, illegal hunting, and illegal fuelwood collecting are documented. Likewise, I explore predation by park wildlife on village crops and livestock and the resulting effects on local livelihoods. I identify a wide range of management responses to these conflicts on the part of both the park administration and local communities and individuals.

The final chapter examines the ways in which Meru peasants contest the state-sanctioned meanings and material uses of Arusha National Park. Based on interviews, local documents, and firsthand observations, it demonstrates how many of the violations of park laws can be understood as efforts to defend or reclaim perceived customary land and resource rights. In particular, what park officials view as relentless encroachment of boundary settlements is revealed as a strategic response by villagers to the insecurity of land tenure historically produced by park policies. I argue that many of the park's policies and the actions of park officials violate the standards of a local moral economy. Ultimately, the portrayal of the national park as pristine nature symbolically and materially appropriates the landscape of Mount Meru for the consumptive pleasures of foreign tourists while denying its human history.

I

Landscapes of Nature, Terrains of Resistance

The peasant's interest in the land expressed through his actions, is incommensurate with scenic landscape. Most, (not all) European landscape painting was addressed to a visitor from the city, later called a tourist; the landscape is his view, the splendor of it is his reward.

John Berger, *About Looking*

Anglo-American Nature Aesthetic

The question of what Africa should look like implies an aesthetic judgment. In the context of national parks, the simple answer is that Africa should look "natural." This is no answer at all as it leads us immediately to the problematic question of what a natural Africa looks like. That is, it leads into a search for an aesthetic definition of nature and thus toward an approach that treats nature and "natural" landscapes as cultural constructions. Crandell's (1993) recent analysis of what she calls the "pictorialization of nature" provides a point of departure. According to Crandell, our definition and assessment of nature (i.e., those of Western-educated societies) are based upon the degree to which it conforms to pictures. Nature became pictorialized, she argues, through a long tradition of artistic representations of landscape which she traces from classical Greece to eighteenth-century England. There, Claude Lorrain's idealized paintings of natural landscapes served as models for the British aristocracy to re-create the pastoral in their estate parks, now often represented as remnants of a distant English past and its origins in nature. This ideal has become such a powerful force in our worldview that "when we think of nature we too often conjure up images borrowed from eighteenth-century England" (8), a consciously designed landscape.

Much of Crandell's analysis covers familiar ground. Raymond Williams wrote two decades earlier: "The main argument is well known. Eighteenth-century landlords, going on the Grand Tour and collecting their pictures by Claude and Poussin, learned new ways of looking at landscape and came back to create such landscapes as prospects from their own houses" (1973, 122). Yet Crandell makes an important contribution by emphasizing how deeply pictorialized nature is embedded in Anglo-American culture and that these scenic habits are implicated in the establishment of places like Yellowstone, the world's first national park. Here she leaves us to wonder precisely how the aesthetics of the pastoral ideal influenced Yellowstone, as the association is made but not developed. She implies that the national park ideal is derived from cultural practices long associated with the picturesque. In particular, the aristocratic (later, the middle-class) penchant for excursions into the countryside in search of picturesque landscapes can be seen as a direct precursor to national park tours. Bruce (1994) draws a similar connection between eighteenth-century landscape aesthetics and American national parks, but applies greater emphasis to the idea of the sublime as the basis for picturesque nature. Sublime nature was not merely beautiful nor was it necessarily harmonious. Rather it was constructed around awe-inspiring vastness and grandeur. Sublime nature, associated more with the paintings of Salvator Rosa than Claude, was central to the development of nineteenth-century European romanticism (Cosgrove 1984). Appreciation of sublime nature through excursions to the Alps or the Highlands of the British Isles ultimately became an identifier of middle-class refinement throughout western Europe. Nineteenth-century American romantic constructions of the "wilderness" drew heavily from the English aesthetic tradition of sublime nature. American romanticism eventually played a leading role in the development of the national park and wilderness ideals in the United States (Nash 1982).

It is unnecessary, and in fact misleading, to force a choice between the pastoral or the sublime as the aesthetic basis for the national park ideal. The emergence of the sublime in painting and landscape appreciation was at once a reaction against the contrived pastoral nature of the eighteenth-century estate park and a further step in the cultural evolution of the idea of nature as a source of aesthetic value. As it happened, early national park supporters in the United States drew from both traditions. Furthermore, whether we argue for the pastoral or for the sublime as the primary inspiration for national parks, aesthetics remain central to our understanding of what constitutes nature and its preservation. Pictorialized nature is fundamental in the history of the national park ideal. "Framing" nature in painting, whether pastoral or sublime, transformed it into picturesque scenery, where the observer is placed safely outside of the landscape. Likewise, surveying, bounding, and legally designating a "wild" space makes it accessible for the pleasure and appreciation of world-weary urbanites. The process of designating and managing wilderness is reminiscent of the art of landscaping "that conceals its own artifice" (W. Mitchell 1994, 16; see also Crandell 1993).

If landscape painting and architecture were critical historical forces in shaping European ideas of nature, how were these transferred to colonial Africa and what transformations took place in the process? What role did national parks play? To begin, the lands and peoples of Africa immediately challenged the familiar tropes of nature and landscape of arriving European colonists. The colonial subject was required to "negotiate between two worlds: the recently lost metropolitan home, and the uncoded Otherness of the present" (Bunn 1994, 138). In essence, the colonist was required to construct a transitional landscape form to bridge the gap between metropolitan conventions and colonial reality (1994). Consequently, familiar pictorial codes from the metropole were employed to make sense of an unfamiliar world, which, at the same time, highlighted an estrangement from it (Daniels 1993, 5). Mitchell extends the relationship between landscape and empire even deeper, proposing that "[l]andscape might be seen more profitably as something like the 'dreamwork' of imperialism, unfolding its own movement in time and space from a central point of origin and folding back on itself to disclose images of unresolved ambivalence and unsuppressed resistance" (1994, 10). One "unresolved ambivalence" has been the incompatible European images of Africa as forbidding wasteland or Edenic paradise (see Curtin 1964), which, to a large degree, reflect the more general historical ambivalence toward "wilderness" in Anglo-American culture (Nash 1982). In both of these opposing tropes, Africa is represented as wilderness, a projection of the fifteenth-century European imagination. Under the influence of Classical and Islamic Mediterranean science, European cartographers mapped the globe into spheres of civilization and savagery (or, alternatively, decadence and innocence) (Cosgrove 1995). It is principally the vision of Africa as earthly Eden—a romanticized wilderness in opposition to the decadent metropole—that underpins the historical development of the national park ideal in colonial Africa (Anderson and Grove 1987; Grove 1990; MacKenzie 1990; Neumann 1995b; 1996). The Edenic myth carried the greatest currency in the settler colonies of East and southern Africa: "Here, at its crudest, Africa has been portrayed as offering the opportunity to experience a wild and natural environment which was no longer available in the domesticated landscapes of Europe" (Anderson and Grove 1987, 4). Colonists were compelled to make the new landscape their own, to employ familiar visual idioms in the construction of a coherent national identity at once separate from the colonized Other, yet not wholly dependent on the metropolitan landscape they had abandoned. In short, imperial nationalists sought "to annex the home-lands of others in their identity myths" (Daniels 1993, 5).

The identity myth of a colonizing society returning to or discovering an earthly Eden is deeply implicated in the establishment of national parks. Significantly, the Edenic vision of the landscape was capable of accommodating an African presence, because incorporated in the Eden myth is the myth of the noble savage. The noble savage, being closer to nature than civilization, could, hypothetically, be protected as a vital part of the natural landscape. To protect the primitive, the savage, was to preserve something of Europe's own origins, a remnant of the natural state of humanity that we gave up to take the path of civilization and cultural advancement (Torgovnick 1990, 46). George Catlin, the American landscape painter credited with inventing the term *national park* (Nash 1982), had included Native Americans in his vision of nature preservation in the western United States. The primitive is picturesque. In dialectical relation to the identity myth of the civilized European, the myth of the Other as noble savage initially could be accommodated within national parks. European settlers with educated sensibilities could admire native innocence and "naturalness" (once initial conquest was completed) as part of an overall exercise in the appreciation of nature. Africans, in other words, could be preserved as part of the natural fauna (Neumann 1995b). In a later chapter, we will see that national parks in Tanzania could accommodate the presence of the noble savage for only a brief time.

National parks provide one important vision of what Africa "should" look like. We have seen that what is being represented is a culturally and socially constructed natural African landscape, and that this representation developed as part of the expansion of European empire. But I am beginning to run slightly ahead of my argument. I still must explain what constitutes a European vision of a natural landscape and its historical and political context. Also, I have yet to fully investigate the idea of nature as a cultural construction. In order to elaborate on the nature aesthetic that I have argued is crucial to the development of the national park ideal, I must further explore the historical development of the landscape idea and how it relates to our understanding of nature.

LANDSCAPES OF PRODUCTION, LANDSCAPES OF CONSUMPTION

The fundamental change in property relations resulting from the transition to capitalist forms of production in Europe was accompanied by sweeping changes in consciousness toward land, as reflected in painting, literature, and landscape architecture. Landscape painting, in particular, was capable of powerfully depicting and shaping ideas about land, nature, and society. The widely accepted argument (cf. W. Mitchell 1994) for the emergence of a uniquely European style of landscape painting runs as follows. The development of the genre can be traced to the urban centers of mercantile capitalism of fifteenth-century Flanders and upper Italy (Cosgrove 1984; Crandell 1993). The invention of perspective around this time was critical to the new art form, allowing the landscape backgrounds of portraits and religious paintings to be extended into the distance and, eventually, supporting the artists' and their commissioners' claims of realism. The position that landscape paintings accurately depicted "reality" was an ideological one. The reality being portrayed often legitimated political authority, reinforced ideas of individual private property, and projected a particular vision of rural social relations (Barrell 1980; Cosgrove 1984; Pugh 1990). "Realistic" landscape portrayal was often the result of painters systematically ignoring fundamental changes in the land (Prince 1988). By the time the genre had fully developed in the nineteenth century, landscape painting was capable of producing enduring mythologies of place and nature, rich in imagery and symbolism, yet convincingly realistic. Edwin Landseer's paintings of the Scottish Highlands for Queen Victoria, to take one case, served to create and reinforce a landscape myth that denies both history and the contingency of place (Pringle 1988). Rural life in Victoria's reign was an important refuge for royalty, and the queen was clear in her instructions to the artist about how she wanted the royal presence in the Scottish Highlands to be represented (ibid.). The result of the royal commissions in Scotland is a series of paintings that help construct a myth of a tranquil landscape filled with loyal subjects, thereby obfuscating the social upheaval of the Highlands' economic transformation.

A central feature in the representation of social relations in landscape paintings was the portrayal of labor. For example, the portrayal of the rural poor in paintings went through several transformations from the eighteenth to the nineteenth century, but in nearly all instances the tendency was either to hide the poor in background shadows or stress cheerful, contented laborers working in harmony with nature (Barrell 1980). Frequently in European landscape painting, the representation of labor was most notable in its absence. T. J. Clark's careful study of Claude Monet's work, for example, demonstrated that "[i]ndustry can be represented, but not labor" in the popular landscape paintings of Argenteuil outside of Paris (1984, 189). The removal of labor from the landscape transformed the world into a series of vistas or "pleasing prospects" (R. Williams 1973, 120) for the pleasure of the outside observer, allowing nature to "become the whole subject of a picture and from which certain classes of human affairs, the practical as opposed to the contemplative, could be excluded" (J. B. Jackson quoted in Cosgrove 1984, 34). As Raymond Williams said, in his much-quoted passage, "[A] working country is hardly ever a landscape. The very idea of landscape implies separation and observation" (1973, 120).

The representation and appreciation of landscape, then, initiated the development of an insider-outsider duality, where the outsider commanded all-encompassing visual control over a nature free of human labor. Developments in transportation technology conspired to strengthen this duality and the equation of nature with the picturesque: "Nature was now experienced as a scenic view gliding past the window of the train or the porthole of the steamer. From these observation posts the landscape looked rather like a framed painting or a photograph" (Frykman and Lofgren 1987, 55). One traveled *through* the landscape as an observer "taking in" (consuming) the scenery, rather than traveled *in* the landscape. In contrast, for the insider, there is no firm distinction between herself or himself and the land, no way to simply step out of the picture or the landscape. Presumably, insiders (e.g., peasants) are not susceptible to the aesthetic appeal of the land-scape as an objet d'art, as Berger (1980) intimates in the epigraph to this chapter.

Cosgrove (1984) has argued that this characteristic separation of the observer from the object of observation, over which the observer is given a sense of ownership and control, constitutes a "way of seeing" the landscape that developed first within the culture of European classes of landowners and capitalists. The emergence and evolution of the land-scape way of seeing can only be understood within the context of the massive social changes in the transition to industrial capitalism. As Williams proposed, one way in which the transition was manifested in European culture was the division of the landscape into two spheres, practical and aesthetic: "[T]he moment came when a different kind of observer felt he must divide these observations into 'practical' and 'aes-thetic', and if he did this with sufficient confidence he could deny to all his predecessors what he then described, in himself, as 'elevated sensi-bility'. The point is not so much that he made this division. It is that he needed and was in a position to do it, and that this need and position are parts of a social history, in the separation of production and con-sumption" (1973, 121). Similarly, in writing about the relatively rapid transition to industrial capitalism in Sweden in the late nineteenth cen-tury, ethnographers Frykman and Lofgren (1987) describe two types of landscapes associated with the new division of labor. The productive landscape is ruled by "rationality [and] profit," the consumptive land-scape by "recreation [and] contemplation" (51).

With the advancement of industrial capitalism in England in the nineteenth century and the increasing use of labor for the production of exchange value, the landscape ideal gradually lost its power to por-tray idealized social relations. According to Cosgrove, "If cultivated land, resources and labour were increasingly unnatural, nature could only exist where human society had not intervened, or at least where the appearance of non-intervention could be sustained, in the wild and unused parts of the environment" (1984, 232). By the end of the nine-teenth century, landscape and nature had become almost interchange-able categories. Idealized "nature" came to bear ever greater moral and ideological weight and pristine landscapes in the countryside were made virtuous in opposition to the polluted cities crowded with workers (Bar-rell 1980; Daniels and Cosgrove 1988).

Parallel to their spatial separation, production and consumption began to occupy distinct temporal spheres of work time (production) and leisure time (consumption). With the emergence of a working class and a petite bourgeoisie, plus opportunities for travel afforded by new transportation technologies, leisure became a mass phenomenon. It was an industry dependent upon the existence of picturesque landscapes. As Edward Thompson explained, "In mature capitalist society all time must be consumed, marketed, put to *use;* it is offensive for the labour force merely to 'pass the time'" (1967, 91; emphasis in the original). One result was the development of a cartographic convention that, in Lefebvre's acerbic words, "designates places where a ravenous consumption picks over the last remnants of nature and of the past in search of whatever nourishment my be obtained from the *signs* of anything historical or original" (Lefebvre 1991, 84; emphasis in the original).

These "remnants of nature" were called upon to play an expanded symbolic role as new social classes proliferated under industrial capitalism. The cultured appreciation of landscape became a mark of "elevated sensibility" that only the "proper" (bourgeois) combination of breeding and education could produce (R. Williams 1973; Clark 1984). Thus it was within and around "landscapes of consumption" (see Williams 1973; Frykman and Lofgren 1987; Cosgrove 1984) that an important aspect of the struggle over class identity, and the cultural values that served to define class, took place. We can find this interclass struggle over the use of leisure time repeated in different periods and in varying locations, but always at the historical point when emergent classes were in search of their own unique identity, and when an established bourgeoisie was pressed to keep itself above the mob and in control of cultural production. In late-nineteenth- and early-twentieth-century Sweden, for instance, the newly established working class found itself in conflict over the ways in which this new leisure time was spent. With the increased availability of mass transportation, higher wages, and shorter work weeks, the countryside and seaside became available to the working class. Many middle-class observers were put off by these developments, complaining that the problem with "working-class vacationers was that they did not know how to behave" (Frykman and Lofgren 1987, 72). By the late eighteenth century in France, the right to experience and appreciate nature undisturbed by insensitive crowds was central to class identity: "There was a struggle being waged in these decades for the right to bourgeois identity . . . [T]he claim to pleasure was nothing if not an attempt to have access to nature . . . To have access to Nature be the test of class is to shift the argument to usefully irrefutable ground: the bourgeoisie's Nature is not unlike the aristocracy's Blood: what the false bourgeois has is false nature" (Clark 1984, 155–56). Thus the moral and cultural superiority of certain social classes was constructed, in part, on the foundation of a refined, aesthetic appreciation of nature.

The consumption of nature thus played an important symbolic role during a period of fundamental change in economic and social life. By the late nineteenth century, nature appreciation had become an index of education and good breeding. The cultural values associated with nature appreciation and its importance for class identity, however, by no means disappeared once the transition was complete. In 1955, a speaker at the Royal Society of Arts in London groped to explain the intangibles that make the appreciation of nature and landscapes incomprehensible to members of the lower classes: "The man-in-the-street[,] . . . however you educate him, . . . will never be interested. The chap who is interested is somebody with a particular form of mentality which has grown up with him through tradition, through inheritance and lots of other things. . . . We shall not go anywhere until we can get back to the old idea of the Manor" (J. L. P. Macnair, quoted in Lowenthal and Prince 1964, 326). We will continue to hear echoes of these sentiments reverberating in the discourse of nature preservationists in North America and, later, Africa, where in an imperial setting the struggle over identity, mediated through encounters with nature, becomes less a matter of class and more one of race and nationalism.

NATIONAL PARKS AS LANDSCAPES OF CONSUMPTION

Understanding the history of the concept of landscape and its role in the cultural constructions of nature is critical to understanding how the national park ideal has developed. The historical point (roughly, the mid-nineteenth century) when the landscape ideal was most fully developed in England and North America, and the concepts of landscape and nature had become increasingly interchangeable, was precisely the point at which the national park movement emerged. National parks in North America were given unique meaning and form different from those of English landscape parks, but the aesthetic ideals that underpinned the movement were imported from Europe. We should not be surprised, then, that the impetus for nature protection came from Europeans and residents of the eastern United States cities "of literary and artistic bents" (Nash 1982, 96). As noted above, the painter George Catlin was, in 1832, the first to use the term "a nation's park." The paintings of Yellowstone by another landscape artist, Thomas Moran, were central to the campaign to establish it as a national park. Landscape painters like Thomas Cole adopted the painterly conventions of Claude Lorrain and the poetry of Wordsworth to project a romantic image of an American wilderness that greatly fueled the national parks movement. Frederick Law Olmsted, landscape architect and one of the first commissioners of Yosemite—an area designated as a park under California state jurisdiction in 1864, eight years before Yellowstone's designation—felt that Englishman William Gilpin's descriptions of "picturesque" tours of the British Isles were essential reading for his employees (Crandell 1993, 132). Olmsted valued the importance of nature appreciation in the constitution of bourgeois identity, reasoning that "the power of scenery to affect men is, in a large way, proportionate to the degree of their civilization and the degree in which their taste has been cultivated" (quoted in Nash 1982, 106). Thus, in the development of the national park concept, we witness a convergence of ideas about landscape appreciation, social identity, and nature protection.

The critical component, symbolically and practically, of the law that created Yellowstone as a model for all of the world's national parks was the prohibition against human settlement and activities. This clause reflects the essence of the landscape way of seeing: the removal of all evidence of human labor, the separation of the observer from the land, and the spatial division of production and consumption. *A national park is the quintessential landscape of consumption for modern society.* At the time of its establishment, Yellowstone was foremost a "pleasuring ground" (the term used in the original legislation), receiving strong backing in Congress from the railroad industry, which saw in it the potential to increase ridership through mass tourism. Hence, the spatial division of consumption from production does not mean that landscapes of consumption are not profitable: "The truth is that all this seemingly non-productive expanse is planned with the greatest care: centralized, organized, hierarchized, symbolized and programmed to the nth degree, it serves the interests of the tour-operators, bankers and entrepreneurs of places such as London and Hamburg" (Lefebvre 1991, 59). As the Yellowstone model has spread, particularly in the tropical Third World, it has provided one anchor for the expansion of global tourism, which is destined to soon become the world's largest industry. Thus, from its inception, the national park has been the principal site where nature is marketed for mass consumption by an increasingly mobile and urban society.

None of these remarks should be interpreted as a denial of the ecological importance that national parks have now acquired. Their roles as in situ gene banks and endangered species refuges are, in effect, the more recent layers of meaning attached to parks. When the park concept was evolving in the nineteenth century, the word "ecology" had just been coined, and familiar ecological terms such as "consumer" or "producer" were still restricted to the realm of economics (Worster 1985). Ecological rationalizations for park establishment have multiplied tremendously since Yellowstone was created. Tropical deforestation and species extinction have increasingly become major motivations for establishing protected areas. These current variations in the meaning and purpose of national parks, however, do not diminish the importance of an Anglo-American nature aesthetic to the origins of the concept and to continued policy and management decisions.

Theorizing Nature

My characterization of the national park ideal as an expression of a culturally and historically contingent nature aesthetic raises critical epistemological and ontological questions about nature whose explication lies far beyond the scope of this book, let alone this chapter. Nevertheless, a degree of transparency in argumentation is called for, since any discussion of nature "protection" in the form of national parks must be based upon an explicitly acknowledged understanding of the relationship between nature and society. The central problematic immediately faced by any investigation of society-nature relations is the fact that it has often been framed dualistically. That is, human society is seen to exist separate from an external nature that can be "dominated," "conquered," or "protected." Recent writings in cultural geography and cultural studies (many of which were cited in the above discussion) have attempted to problematize this conceptualization of nature as external to human society. Rejecting the concept of landscape as object, these "new" cultural geographers (Jackson 1989, 1) "prefer metaphors of icon, spectacle, way of seeing, or theater" (Demeritt 1994, 164) in order to explore how nature and landscape are actively constructed through social practice and imbued with meaning through cultural representations (see, e.g., Cosgrove 1984; Cosgrove and Daniels 1988; Jackson 1989; Daniels 1993).

North American environmental historians have also been grappling with questions about society-nature relations (e.g., Crosby 1986; Cronon 1983; Worster 1985; 1990; and Merchant 1980; 1990). Much of their concern has been with reintroducing nature as an active force in human history, a project that some have argued puts them at odds with the writings of the above-cited cultural geographers (see Demeritt 1994). Yet environmental historians hardly present a united conceptualization of nature that can be placed in a straightforward opposition to cultural geographers. Recent theorizing on the relationship between nature and society has revealed deep ideological rifts in the new field.[1] Elizabeth Bird, for example, clearly establishes her position "that scientific knowledge should not be regarded as a *representation* of nature, but rather a socially constructed interpretation with an already socially constructed natural-technical object of inquiry" (1987, 255; emphasis in the original). She includes in her camp colleagues Donald Worster and Carolyn Merchant. While Merchant's (1990) attention to changes in consciousness toward nature in relation to changes in production practices coincides with the notion of socially produced nature, Worster's approach is quite the opposite. Worster (1990) does not see nature as merely culturally constituted or socially produced but rather as an external object, presenting a theoretical hierarchy wherein "nature" forms the *base*, the mode of production constitutes the *structure*, and "ideology" the *superstructure*. Similarly, Weiskel takes nature as an objective given that can be used as a standard against which we can gauge a society's adaptive success (1987).

Within geography, the failure to reach an agreement on the question of nature as an external objective reality or as a cultural construction is at the heart of the schism between the human and physical domains (FitzSimmons 1989). In the past decade or so, several Marxist geographers have attempted to address the question of society-nature relations and incorporate environment into the study of society (Burgess 1978; Smith and O'Keefe 1980; Smith 1984, Redclift 1987; FitzSimmons 1989). Perhaps the most fully developed of these efforts has been Neil Smith's *Uneven Development* (1984). Beginning with a critique of positivist views of external nature, Smith argued that the way out of the dualism is through Marx's ideas on the dialectical unity of nature and society—nature is shaped by human labor, and the laborer is in turn shaped through this encounter. In answer to this dualism, Smith proposes two concepts: "first nature," that which is unaltered by humans; and "second nature," which is the institutions—the market, the state, money—that have developed to regulate commodity exchange. Under capitalism, "first nature" disappears and becomes just another product, even in the cases of supposedly pristine nature in national parks: "These are produced environments in every conceivable sense. From the management of wildlife to the alteration of the landscape by human occupancy, the material environment bears the stamp of human labor; from the beauty salons to the restaurants, and from the camper parks to the Yogi Bear postcards, Yosemite and Yellowstone are neatly packaged cultural experiences of environment on which substantial profits are recorded each year" (57). This is not to say that capitalism is responsible for the disappearance of pristine nature. Recently, historical and cultural geographers—most of them not engaged in social theory or critiques of capitalism—have clearly demonstrated the ancient role of human agency in shaping what was once considered "wilderness," and in the process have called into question the very idea of pristine nature. A recent example is the fine 1992 special issue of the *Annals of the Association of American Geographers* on the population and landscape of the pre-Colombian Americas. What is unique about the human transformation of nature under capitalism is that it is now part of a global system within which even the most geographically remote regions are incorporated as part of the process of uneven development (Smith 1984).

Smith's comment on the marketing of national parks echoes the words of Lefebvre cited above. Noting how nature (in the form of national parks) is packaged by capitalism for mass consumption, Lefebvre recognized that "it is not at all easy to decide whether such places are natural or artificial" (Lefebvre 1991, 83). This, of course, was precisely the goal of eighteenth-century landscape architects working on estates of the English aristocracy—to "naturalize" the parks and hide all evidence of their artifice. In the case of national parks, every effort is made to disguise or ignore the evidence of human agency, either historically or in the present. This denial of human agency in the popular representations of nature, observes Raymond Williams, is a source of confusion surrounding the search for pristine nature: "Some forms of this popular modern idea of nature seem to me to depend on a suppression of the history of human labour, and the fact that they are often in conflict with what is seen as the exploitation or destruction of nature may in the end be less important than the no less certain fact that they often confuse us about what nature and the natural are and might be"

(1980, 78). In sum, the denial of human history and agency is implicit in the idea of nature that has developed for mass consumption under capitalism.

Where does this leave us in regard to the question of nature as historical agent versus nature as socially produced and culturally constructed? To begin, contra Demeritt (1994), I do not see these conceptualizations of nature as mutually exclusive. Redclift, following Andrew Sayer (1983), states that we can theorize how biological powers are mediated by social forms, "without wishing to argue that they are reducible to social forms alone. The environment is more than the production of Nature" (1987, 229). Neither Smith (1984) nor Lefebvre (1991) would contest this position. Smith explains that though society may incorporate nature, it does "not cease to be natural in the sense that they [the elements of first nature] are somehow now immune from non-human forces and processes—gravity, physical pressure, chemical transformation, biological interaction" (47). Thus, I hope to show that while the representation of nature in national parks can be traced to an Anglo-American nature aesthetic and to the creation under capitalism of landscapes of consumption for leisure and profit, natural processes continue to operate, sometimes in ways that challenge or contradict the preservationist aims of some human interventions. Hence the need for bureaucratized activities bearing such oxymoronic labels as "wilderness management" and "wildlife management." Nature surfaces as an actor repeatedly and in unexpected ways to thwart our most diligent efforts to "preserve" it. In other words, national parks are paradoxical. The culturally constructed aesthetic ideal of the natural landscape can never be "preserved" because the dynamism of ecological processes defies preservation.

Alston Chase (1987) has documented at length some of these contradictions in the history of Yellowstone National Park. I want to present one example—the decades-long management effort to suppress fire—and in the process bring us full circle to the importance of nature aesthetics in the national park ideal. Fire was long considered anathema in national parks because it destroyed the picturesque and challenged the notion of nature "preservation." Its suppression, however, created unexpected and "undesirable" effects in nature. After decades without fire, dead vegetation had accumulated to such a level that, under favorable weather conditions, suppression of wildfire was virtually impossible. What certainly ranks as one of the largest and most pervasive propaganda campaigns of the United States government, that of "Smokey the Bear," thus had to be radically reconceived. The new campaign to reeducate park visitors about the ecological importance of fire—its "naturalness"—hardly matches the original campaign in scope and effectiveness. However ecologically sound or "natural," no one is prepared to conclude that expanses of burned-out stumps have become a popular tourist attraction. This brief example illustrates that, in the case of national parks, nature functions as an active agent, while simultaneously being socially produced. The conditions that led to uncontrollable wildfires in Yellowstone were socially produced as part of marketing nature for mass consumption. Removing the human or natural agents in the landscape in order to preserve nature creates a contradiction in that ecological change begins immediately in response and often management intervention is required to slow or control change. It also demonstrates how the marketing of national park tourism and the continuing dilemma over fire management in parks can be traced to the fact that we have been and continue to be beholden to an Anglo-American aesthetic of the way nature should look.

Landscape, Nature, and Conquest

IMAGINING A "NATURAL" HISTORY

When we deny—whether in naturalistic paintings or national parks—the role of the human hand in shaping landscape, we contribute to the validation of a particular historical narrative of European imperialism. Within this narrative, colonialism and the global expansion of European economic and cultural hegemony involved the transformation of vacant wilderness into productive fields, factories, and cities.[2] Within this model of history, national parks represent remnants of the pre-European landscape, pockets of remote, unoccupied wildlands preserved as reminders of a "national heritage." While geographers and environmental historians have thoroughly debunked the notion of pristine nature in most terrestrial landscapes, it remains a popular idealization, with national parks serving as the living model. Hecht and Cockburn recently challenged the notion of national parks as vacant lands in an appendix to their book, *Fate of the Forest*. In the process, they demonstrated that the history of Anglo-American conquest of Native Americans and the history of national parks are inextricably linked.

They detail the Indian wars against a small group of Miwok warriors whose defeat by the United States military cleared the way for the establishment of Yosemite Valley as a tourist attraction. The land that was eventually designated a national park was "Miwok hunting and collecting grounds that bore the imprint of their fires, their planting, and their husbandry for thousands of years" (Hecht and Cockburn 1990, 274). Chase makes a similar argument in regard to the establishment of Yellowstone National Park: "[N]o sooner was the park created than the Indian . . . was forgotten . . . In nearly all service publications for a century or more, Indians, when mentioned at all are called early 'visitors' to Yellowstone . . . 'Yellowstone was virtually uninhabited in 1872,' wrote former Yellowstone naturalist Paul Schullery in 1984" (1987, 107). Chase takes issue with the government's version of history, claiming that Shoshone, Crow, Bannock, Blackfoot and, above all, Sheepeater peoples had all occupied and utilized the lands of Yellowstone, putting their mark on the landscape over thousands of years. Warren (1994) recently documented Bannock hunters' losing struggle with Euro-American commercial hunters and the United States government over access and control of wildlife in the greater Yellowstone area.

All of these accounts illustrate how the histories of national park landscapes are now openly contested and increasingly politicized. Frequently, however, challenges are neutralized, or submerged, because history is often written by society's elites with vested interests in projecting certain ideals and values onto the past (Scott 1985). Local voices that might contest the accepted story of nature protection are suppressed, principally because of their lack of access to the institutions that generate historiographies or, as in the case of the Miwok, because they have been permanently silenced. The vortex around which national park histories spin is the black hole of the "natural landscape," into which the imprint of the human hand disappears. It is not enough to physically remove human agency and occupation from the landscape, they must be purged from history completely. We can begin to see how national parks have been looked to in North America (and elsewhere, as we will see below) to carry a heavy burden of symbolic meaning. In North America, the national parks were intended to, among other things, preserve a sample of the "national heritage"; that is, to preserve the memory of an idealized pioneer history as an encounter with "wilderness" that was conquered by enterprising Europeans. With the aid of national parks, the history of the conquest of humans—the Native American societies that occupied the continent for thousands of years—was transformed into the conquest of nature. Parks help to conceal the violence of conquest and in so doing not only deny the Other their history, but also create a new history in which the Other literally has no place. As Raymond Williams wrote of England's estate parks, the achievement is "an effective and still imposing mystification" (1973, 125). National parks, as representations of a harmonious, untouched space of nature, mask the colonial dislocations and obliterate the history of those dislocations, along with the history of the spaces that existed previously.

The ideals and values attached to the North American frontier have by no means lost their hold on American political consciousness. Witness the repercussions of the Smithsonian Institution's 1991 National Museum of American Art exhibit, *The West as America: Reinterpreting Images of the Frontier, 1820–1920*. The exhibit was designed to situate paintings of the nineteenth-century American landscape within their social-historical context. That is, the exhibit in part sought to demonstrate that a complex process of European expansion involved not just encounters with nature but with a multicultural mix of African Americans, Native Americans, and Chinese. The exhibit revealed how the painting of the frontier "conspired to convince Americans they could find themselves in the untamed West as they embarked upon a civilizing mission" (Pudup and Watts 1991, 11). Politicians from western states in particular were quick to take issue with the exhibit's reinterpretations. Senator Ted Stevens of Alaska made it a national issue, accusing "the museum of advancing a leftist political agenda" and threatening to cut the Smithsonian's budget appropriation (Foner and Wiener 1991, 163). "Perverse" was the word chosen to describe the exhibit by Stevens and others, and the exhibit quickly faded away when museums in Saint Louis and Denver canceled their bookings (Pudup and Watts 1991).

The Smithsonian exhibit did not create a political stir because it was excessively radical in its efforts to draw connections between "artistic patronage and historical representations, [and] between myth and ideology in art" (10–11). The real problem arose because the exhibit, by critiquing visual representations of the West, threw into question an already fragile national identity based on the myth of wilderness conquest and the European civilizing mission.

In Africa, national parks are being subjected to similar critiques, particularly in countries only recently liberated from minority European settler control, such as Zimbabwe and South Africa. Parks in these countries are the topic of political debate precisely because of their symbolic importance for the construction of a national identity for

European settlers, now rendered increasingly fragile by the successes of African liberation movements. Caruthers (1989) explains that South Africa's establishment of Kruger National Park in 1926 came at a critical juncture in the political development of the republic, serving as an important unifying symbol for white national identity. As in North America, the park was represented as a remnant of the African "wilds" that the Afrikaners had struggled to tame. Naming the park after Paul Kruger was a calculated decision designed to enlist the support of Afrikaners and create "a symbol of cultural unity" (1994, 272) for whites. Similarly in Zimbabwe, the creation of Matopos National Park in 1926 began decades of political struggle around two competing myths of the Matopos Hills, one white and one black (Ranger 1989). The fulcrum of the struggle was the "different ideas of 'heritage' and how best to preserve it" (218). At the height of the guerrilla movement in the 1970s, the national park became the focus of a political battle between two competing myths. Whites emphasized the natural heritage of the park, and the guerrillas and their supporters emphasized black cultural rights: a natural history versus a social history. Once again, the settler version of history embodied in the park supports the myth of wilderness conquest. As a result, Matopos at times became "emblematic of the wider conflict between black and white" (242).

The historical cases of North America and Africa suggest the importance of national parks to the formation of a national identity for the dominant settler culture, an identity forged through a mythologized encounter with nature. National parks have, as W. J. Mitchell proposed for imperial landscape painting, a "historical function in the formation of a colonial national identity" (1994, 23). European representations of landscape in colonial settings may be seen as part of the "consequent displacement of the country-city dichotomy onto world geography" (Bunn 1994, 129). The colonies were viewed by Europeans metaphorically as the countryside (Said 1994; Pugh 1990). In the imperial European conceptual map of the world, Europe was culture and the colonies were nature. The discoveries of vast forests and plains inhabited by unfamiliar peoples led to wild speculations in popular and scholarly writings in Europe that Eden had been rediscovered. Occupied by the race of Adam, the Americas were believed to be "undisturbed since the creation" (Glacken 1967, 686). Early nationalist sentiments in the United States often offered North America's "natural beauties" and "raw nature" as a heritage to be treasured like the cultural achievements of Europe's cities (Nash 1982). As Smith explains, "Where the dominant social symbols of the Old World drew their strength and legitimacy from history, New World symbols were more likely to invest in nature" (1984, 7). The countries mentioned above in which parks are important rallying symbols for national identity are *settler states of European colonial expansion.* The national identity being fought over is that of an immigrant population whose history has been left behind in the Old World and who must forge a new identity out of the conquest of nature. In this context, parks are but one aspect of the portrayal of global history as the history of Europeans, with the rest of humanity frozen in some primeval state (Wolf 1982; Blaut 1993). Hence, an important role of national parks in the construction of class and racial identity is to eliminate the record of indigenous history and culture, replacing it with a vacant landscape into which Europeans streamed.

Fifteen years after publication of his highly regarded *Wilderness and the American Mind*, Nash added a new chapter, "The International Perspective," to the third edition (1982), within which he speculates on the internationalization of nature protection, particularly the global expansion of the national park ideal. He divides the world's countries into "nature importers," societies where urbanization and industrialization have led to the increased valuation of undisturbed nature, and "nature exporters," countries less advanced on the "development" trajectory and which therefore have an overabundance of nature (346). Hence whites in Africa who came from nature-importing countries were able to recognize the value of "the last unmodified African environments" (356). According to Nash, "Colonization made it easy" (354) for these visionaries to work for the preservation of these remnants of nature. In a more recently published explanation, Harmon attributes the rapid establishment of national parks worldwide to the "demonstration effect" (1987, 151). The power of the demonstration effect is particularly strong in "developing countries," where the "exposure to the national park ideal has proven irresistible to the governments" (ibid.).

Implicit in Harmon's thesis is the unquestioned superiority of "Western" ideals, values, and beliefs. That is, the "Third World" need only be exposed to the modern world to become dissatisfied with "traditional" ways and eventually abandon their own culture for the West's. It is, in essence, a variation of a widespread notion of history that geographer J. M. Blaut labels "Eurocentric diffusionism" (1993, 1)—the idea that cultural innovation tends to flow one way, from Europe to the rest of the globe. In Nash, we can quickly recognize the colonial geography of a world divided into regions of nature and regions of culture. With this

model as a foundation, he posits a moral struggle by foresighted individuals emerging to save the last remnants of Eden. Preservation
activists in Africa have long used this argument: "Europe has its *cathe-
drals*, preserved through ages; Africa is proud to show the prodigious
natural spectacles which she has helped to save" (J. Verschuren of the
Congo National Parks Institute, quoted in Watterson 1963, 55; emphasis added). This sentiment, labeled the "Eden complex" by Anderson
and Grove (1987), was common among European preservationists and,
after World War II, expressed increasingly in the terminology of the new
science of ecology. The first director general of UNESCO, for example,
appealed for the creation of parks in eastern Africa because it "contains
the last accessible portions of the prehuman world's climax community" (Huxley 1961, 79). The conservationists' version of history presented here is predicated on an explicit denial of the existence of
preconquest societies, thereby portraying a deeply political struggle (as
we shall see in the case of Tanzania) as a moral one (see Anderson and
Grove 1987).

CONQUEST, CONSERVATION,
AND PROPERTY RELATIONS

The politics of national parks in colonial Africa were not
only confined to the use of the parks as symbols of white national identity. National parks (and conservation laws in general) were one component of the broader process of colonial appropriation of land and natural resources. In South Africa, the establishment of Kruger National
Park was "part of the process of the systematic domination of Africans
by whites," constituting "yet another strand in . . . the struggle
between black and white over land and labour" (Caruthers 1989, 189).
Most of the lands upon which European colonial authorities established
their national parks had either been cleared of people in the course of
conquest or, as in the cases of Kruger and Matopos, alienated from
Africans after their legal designation. These dislocations, combined with
the development and dominance of state-controlled, scientific natural
resource management, severely disrupted the production and reproduction strategies of rural African households.[3] Under colonial conservation laws, the collection of fuelwood became wood theft, the hunting
of animals became poaching, and pasturing cattle became grazing trespass, with all the ramifications for state violence that these meanings
imply (Neumann 1995b). National park and conservation laws are at

once one part of the wider process of land and resources seizure by the colonial state, and a symbolic legitimation of that process.

We can trace a great deal of the impetus for the dislocations of African settlements and claims, and the criminalization of customary practices in the name of wildlife conservation, to the lobbying of colonial governments by well-to-do white hunters (MacKenzie 1988; Neumann 1996). In British-ruled Africa, this lobbying was conducted through the London-based Society for the Preservation of the Fauna of the Empire (SPFE). I will elaborate the history of the SPFE's intervention in colonial conservation policy in later chapters. For now, I merely want to tie together the above discussion of national parks with the colonial state's generalized seizure of African lands and natural resources. According to the SPFE, white "sport" hunting and nature preservation went hand in hand, while traditional African hunting was antithetical to preservation (Neumann 1996). They successfully lobbied the Colonial Office to create systems of game reserves and national parks within the colonies where most African customary claims were voided, and helped write the colonial game laws that protected white privilege while greatly restricting African hunting (ibid.). The state's authority to do this came from the British government's claim that all the land and natural resources within colonial boundaries were legally the property of the Crown. Thus the game reserves, which provided the core lands of many future national parks, were part of the shift from local, customary property rights to state control.

European conquest in Africa, then, meant a dramatic restructuring of the property relations. Hunting laws and the creation of game reserves and national parks were one component of this process. The symbolic uses of nature in Africa to legitimate political claims and portray the values of a particular culture or class as universal are reminiscent of the ways in which landscape paintings and private estate parks in England were employed during the capitalist transition. Indeed, there are many parallels between the African dislocations and loss of customary claims and the historical situation of rural England (Neumann 1996). Raymond Williams's observation that private estate parks in eighteenth-century England "were the formal declaration of where the power now lay" (1973, 107) could be made for national parks in colonial Africa, appropriated for the exclusive use of whites.

Within African agrarian societies, parks have affected the meanings of land and customary land and resources uses in much the same way as England's enclosure acts, which promoted private property rights at the expense of communal rights during the transition to capitalism. From the early eighteenth to the late nineteenth century the thousands of enclosure acts, as well as laws like the infamous Black Act, served to eliminate common rights in land and the "products of nature" (1973; Thompson 1975). Enclosure constituted the "legalized seizure" (R. Williams 1973, 98) of a quarter of all cultivated land by a politically dominant capitalist class. This new class, characterized by a greater command over money derived externally from the rural economy, had moved into the forests and pastures, bringing with them a disregard for customary rights and a desire to profit by cashing in the old manorial rights for lands, parks, and money (Thompson 1975). Profits were used to develop estate parks as conspicuous displays of wealth, usually at the expense of customary claims. Whole villages were razed and commons closed to meet aesthetic demands, to create pleasing prospects for the owners and to assert their individual property rights: "Men like Lord Cadogan and Horace Walpole were said to regard estates as commercial enterprises, openly exploiting their tenants and happily removing them wholesale . . . in order to replace their village with a grand prospect of shaven lawns and aesthetically-grouped trees denoting their new found status and power . . ." (Cosgrove 1984, 211). The estate parks, devoid of evidence of human labor, provide the locus for the convergence of social status, political power, nature aesthetics, and restructured property relations.

These changes in social relations in rural England are directly relevant to understanding the mobilization of resistance against new forms of property, and to a generalized pattern of political upheaval in the countryside. In Thompson's (1975) examination of England's early-eighteenth-century "Blacks"—an association of individuals accused of various offenses from illegal hunting to extortion and assault—he sought to uncover the intentions behind their crimes. Thompson's method revolved around developing the larger social context in which the events took place "so that from this context and their actions we can deduce something of their motives" (54). His investigation revealed that the main target of the Blacks was a new class of financiers, merchants, and army officers who had moved into rural England, carving deer parks and estates out of former commons. Thompson concluded that the offenses that the state attributed to the "Blacks" were committed by members of a broad spectrum of rural underclasses acting in defense of customary rights to the game, pastures, and forests that had been privatized and closed off to common access.

The English game laws implemented during the capitalist transition had a devastating effect on peasant society, and violent resistance was widespread. Hopkins (1985), for instance, contends that the most brutal violence from the eighteenth through the nineteenth century was rural, not urban, with much of it attributable to the new game laws. Far from being a period of tranquillity in the countryside, a virtual state of war existed between poachers and gamekeepers who had increasingly harsher penalties to back them up. Each side of the war had its own moral justifications, as landed gentry came to monopolize game hunting. The landless (80 percent of the rural population in the early nineteenth century) were totally deprived of the right to hunt any animals, which further worsened their already marginal condition. As hunting developed as an important identifier of class, game reserves became a major preoccupation (62). The game privilege, so visible a sign of the relationship between property rights and class, was a catalyst for social change. In an effort to get "a bit of [their] own back" (quoted in Hopkins 1985, 265), the people who had lost their customary claims to the privileged classes continued to hunt in defiance of the new property laws that labeled them as "poachers."

Several important points arise from this discussion. First, rights of common access to land and natural resources, whether in colonial Africa or England, have been a central arena of social conflict during the introduction of individual private property in agrarian societies. Second, wildlife and the right to hunt have historically served as powerful symbols of class privilege and an important source of class identity. In colonial Africa, "the Hunt" (MacKenzie 1987), and its association with privilege and status, was an important symbol of European domination and was central to the development of a national parks movement there (Neumann 1996). Third, the introduction of new forms of property in the transition to capitalism in Europe (and, later, in colonial and postcolonial settings) fundamentally violated rural, class-bound assumptions of justice and morality. It is to this last proposition that I now turn my attention.

LANDED MORAL ECONOMIES

The concept of moral economy has been widely applied in a variety of geographic, historical, and social circumstances to explain struggles over norms, values, and expectations related to the livelihoods of subordinate classes during major economic transformations. The current usage of the term originated with Thompson (1971), who had retrieved it from late-eighteenth-century anticapitalist polemicists. His analysis of social conflict surrounding forest access and the motivation of the "Blacks" (Thompson 1975) in eighteenth-century England relies heavily on an assumption of a moral economy of the rural underclasses. The moral economy concept has since been highly contested and debated to the point where Thompson has been critiqued (D. Williams 1984) and has responded (Thompson 1991), and his response has been reviewed (R. Wells 1994). My aim here is to review the problems associated with its sometimes cavalier usage in order to clarify my own engagement with the concept in analyzing struggles over livelihoods and nature preservation in contemporary Tanzania. Since some authors have strayed far from Thompson's original usage and some critics' arguments are based on a misrepresentation of his formulation of moral economy, it is best to begin by reproducing his most recent clarification. He explains that his usage has been "confined to confrontations in the market-place over access (or entitlement) to 'necessities'—essential food. It is not only that there is an identifiable bundle of beliefs, usages and forms associated with the marketing of food in time of dearth, which it is convenient to bind together in a common term, but the deep emotions stirred by dearth, the claims which the crowd made upon the authorities in such crises, and the outrage provoked by profiteering in life-threatening emergencies, imparted a particular 'moral' charge to protest. All of this taken together, is what I understand by moral economy" (1991, 337–38). Thompson's principal historical focus was England in the eighteenth century, where, he argues, prices not wages were the main object of popular protest (1971). From his phenomenally detailed study of food riots, he finds the evidence to reconstruct the "view from below" as a moral economy. Above all, the moral economy of eighteenth-century underclasses mandated "that prices ought, in times of dearth, to be regulated" (112).

Since Thompson's initial use of the term in *The Making of the English Working Class*, historians and social scientists have adopted and used it in quite different ways in a wide variety of social contexts and cultural and historical settings. In a recent review, Roger Wells introduced a useful "demarcation between commercial and landed moral economies" (1994, 278) in the vast and growing moral economy "school" of literature. It is the latter that is most relevant to questions of national parks, land rights, and resource access, and I will restrict the rest of the discussion to those studies that essentially address peasant societies.

In *The Moral Economy of the Peasant,* James Scott (1976) borrowed Thompson's term for use in studying the origins of peasant rebellion in

Southeast Asia. Scott, as Evans notes, was part of a "wave of academic interest in peasant societies" (1987, 193), led by Barrington Moore (1966) and Eric Wolf (1966). These scholars presented what they thought to be key characteristics of precapitalist peasant economies and their related normative systems: first, that every member of the community should have access to a minimum amount of land and resources necessary to meet their subsistence needs and social obligations (B. Moore 1966, 497); second, that moral economies are founded upon social arrangements among community members which help dampen crises of consumption for individual households through a redistribution of surplus (Wolf 1966). Mechanisms for redistribution include access to the commons (B. Moore 1966, 497) and "patterns of reciprocity, forced generosity, communal land and work-sharing" (Scott 1976, 3).

Scott built upon these fundamental characteristics to argue that peasant societies have a "subsistence ethic," and that poor peasants are above all risk-averse. That is, the bottom line for peasants is a right to subsistence, and since access to land in peasant societies is the primary means of achieving subsistence, there exists a set of social arrangements to ensure access. Scott argues that the right to subsistence "as a moral principal . . . forms the standard against which [others'] claims to the surplus . . . are evaluated" (7). For poor peasants, guaranteeing subsistence is far more important than maximizing profits, and so they will tend to engage in arrangements that reduce the risk of falling below subsistence. That is, they will invest in building social ties at all levels, including with wealthy patrons and state officials, based on principles of reciprocity and mutual obligation. It is in the best interests of individuals to build up a variety of relations within their moral community in order to increase their security in many different contexts (Wolf 1966).

The central ideas of landed moral economy have been employed in sub-Saharan Africa as part of a general interest in the continent's perceived agrarian crises. Goran Hyden (1980) has been one of the more influential (and debated) scholars postulating an African "peasant view" of social relations. Hyden uses the term *economy of affection*, rather than *moral economy* but, like Scott (1976, 5), is indebted to Polanyi's theorizing of precapitalist agrarian societies, particularly in his argument that the market economy is historically recent, unique, and fundamentally different from preceding economies wherein social relations were nurtured by systems of reciprocity and redistribution (Polanyi 1957). Thus Hyden emphasizes the importance for peasants of developing affective social ties of mutual obligation in order to expand "their risk-bearing capacity" (1980, 13).[4] In Hyden's view, the political consequence of the economy of affection is a "sometimes active resistance to operating in a framework of impersonal rules" on the part of the peasant.[5] For Hyden, this ability to resist derives from the fact that the peasantry is "uncaptured" by dominant classes in Tanzania. Thus bureaucrats and politicians are entrapped by the peasants in a system of patronage (30).

Watts (1983) evoked the concept of moral economy in explaining the types of practices and norms within Hausa society in northern Nigeria that served to see people through the lean times of drought. In the analysis, he points out the dangers in both Scott (1976) and his main critic, Popkin (1979), of essentializing a stratified, heterogeneous peasantry as either "risk averse" or "rational." Like Scott, however, Watts identifies within the lower Hausa social strata a "safety-first principle" and a "subsistence ethic" (106) that permeated the precapitalist peasant worldview. He also takes pains to not romanticize peasants as more "moral": "[M]oral economy emerges as an outgrowth of class struggles over subsistence minimum and surplus appropriation, not as an attribute of a specific, isolated group" (109). More recently, Steven Feierman described a "subsistence guarantee" within Shambaa society in Tanzania that was threatened by colonial soil erosion control schemes. Feierman very carefully argues that violations of the subsistence guarantee (subsistence ethic) were *the* motivation behind the subsequent resistance to colonial schemes: "[T]he central value being defended was not equality among peasants, nor was it a defense of production for use as opposed to exchange. The central value was peasant welfare . . . the promise of continued subsistence" (1990, 195). Finally, Anderson analyzes cattle raiding and cattle theft among the Kalenjin of colonial Kenya through the lens of moral economy. While making passing reference to Thompson, he appears to equate moral economy with "Kalenjin attitudes towards stock theft" (1986, 400), which revolve around differentiating between community members and outsiders. The Kalenjin, he points out, make a "moral distinction between a raid and a theft," the former conducted against outsiders and the latter against fellow community members.

This brief review will serve as a basis to critically examine the varying ways in which scholars have imported the landed moral economy into Africa. Anderson's 1986 study provides valuable insights into the social meaning of rural crime under colonialism, but his limited and vaguely specified usage of "moral economy" makes it an easy target for critics. That is, he conceptualizes moral economy as synonymous with a local value system or "attitudes." It is precisely this type of usage from which Thompson sought to distance himself, noting that "if values, on their own, make a moral *economy* then we will be turning up moral economies everywhere" (1991, 339; emphasis in the original). To return to Hyden (1980), the political patronage system, which he argues derives from the economy of affection, is difficult to differentiate conceptually from "pork barrel" politics in the contemporary United States. Once again, if class relations mediated by political patronage indicate the existence of a moral economy, we will no doubt find moral economies everywhere.

Casual usage of the term, combined with selective readings of the literature, has allowed for easy criticism, exemplified by *The Rational Peasant* (Popkin 1979), where the moral economy argument is overturned by misrepresenting it (Moise 1982). Timothy Mitchell, while offering a fine critique (discussed below) of recent scholarship emanating from the moral economy "school," also misrepresents the concept. He writes, "The shared theme of these writings is that prior to the triumph of capitalism common people shared an ethic based on reciprocal exchange of gifts and services and redistribution in times of need, rather than individual pursuit of self-interest . . ." (1990, 546–47). None of the moral economy proponents have argued that peasants do not exhibit self-interested behavior, but rather that in a landed moral economy self-interest is pursued through very different types of social relations than under capitalism. Watts (1983), for one, argues that there was plenty of self-interest involved and nothing particularly moral about the moral economy in Hausa society. As Wells indicates, "No historian asserts that moral economy was class-specific, or denies that some plebeians were profit motivated and attracted by upward social mobility" (1994, 281).

If some critics' arrows fall wide of their mark, the concept of moral economy has, nevertheless, inherent limitations that restrict its usefulness for understanding the behavior and motivations of peasants. In a comprehensive review of peasant resistance and protest in Africa, Isaacman notes that the formulation of a localized peasant moral economy in opposition to an external market economy leaves little room for investigations of gender, generational, or intraclass conflicts. There is no peasantry in Africa, but *peasantries* distinguished by specific combinations of labor processes and property relations. He explains that generalizations which assume "that all peasants are essentially 'risk minimizing' [Scott 1976], 'profit maximizing' (Popkin, 1979), or 'uncaptured' (Hyden, 1980) . . . foreclose the possibility of exploring and understanding the very notion of power and politics held by differing peasantries in their specificity" (Isaacman 1990, 21). Evans further argues that a moral economy does not aid us much in uncovering an independent political consciousness among peasants, but simply documents "the 'subordinate value system' of the lower classes" (1987, 211). While I tend to agree that this may often be the case, Thompson insists that the moral economy "is continuously regenerating itself as an anticapitalist critique" (1991, 341).

With these shortcomings in mind, I remain convinced that under certain specified conditions and in particular places and periods of history, a moral economy approach to analyzing the actions and speech of subordinate class members can greatly aid in uncovering motivations, intentions, and politics in everyday peasant life. Models of peasant behavior and normative systems are highly contingent and thus historical, geographical, and social contextualization is essential. I argue that there are situations where threats to livelihood are so pervasive that gender, class, and generational conflicts are submerged and a broader solidarity emerges. Peluso makes this argument for Java's forest villages faced with state laws of resource control. "In facing these 'moral crimes' committed by the state, even a highly differentiated peasantry can mask its class tensions" (Peluso 1992, 11). Thompson's examination of the violence surrounding Hampshire Forest in eighteenth-century England likewise reveals a conflict "between the park owner and the keepers on one side, and most of the rest of the village (including its Rector) on the other" (1975, 227). Similarly, Isaacman and colleagues (1980, 599) describe the suppression of internal community differences in the organization of collective resistance to state-mandated cotton cultivation in colonial Mozambique. The situation that this book addresses, I argue, produces a similar solidarity, though differences do emerge within the communities in the way individuals are affected by and respond to nature protection policies.

Furthermore, I agree completely with the sentiment contained in Roger Wells's statement that, "[u]ltimately, Scott's moral economy is period-specific within the histories of peasant societies, and it is in terms of historical validity that it must be judged" (1994, 294). I would add that my usage is, in addition, place-specific within the geographies of peasant societies. The villages I investigated cannot be thought of as representative of "the Meru" and I do not intend them to be generalizable. Indeed, the ecological, socioeconomic, and cultural variation within the area inhabited by people who call themselves "Meru" renders generalizations about identity meaningless. The villages on the northern park boundary are characterized economically by the absence of widespread cash cropping, wage labor, and a well-developed land market. In the villages on the southern park boundary, these conditions are relatively more prevalent. Both cases, however, differ greatly from "central Meru," where smallholder coffee production and wage labor in Arusha town or its suburbs predominate. Furthermore, the "solidarity" alluded to above is a by-product of nature protection policies, and their impact on local politics fades in proportion to the distance from the park boundary.

In summarizing my understanding and usage of moral economy, I want to explain how it differs from previous investigations. Ultimately I see the situation explicated in this book as much closer to Feierman's (1990) case than Scott's (1976). That is, Scott is concerned with peasants' normative evaluation of surplus extraction in the form of rents and taxes, while Feierman is concerned with the way in which state soil conservation policies—more indirectly related to surplus extraction than rents and taxes—are resisted because they threaten subsistence rights. Nevertheless, my investigation is quite different from the state policies Feierman examined, which essentially were about increasing agricultural production for greater state revenue. In the case of nature protection policies, increased production and surplus extraction are not the central issues. National parks in Tanzania have taken land *out of production,* albeit to generate revenue, but not through direct surplus extraction from rural communities. The cutting off of access to land and resources is thus the source of the tensions surrounding the moral right to subsistence. For villages adjacent to the park boundary, the right to subsistence is further threatened by the relentless destruction of crops by wild animals that, once controlled by local communities (Kjekshus 1977), are now protected by the state to generate revenue from tourism. Under these conditions, customary rights of access that facilitated subsistence are defended as moral rights.

My analysis does not preclude the potential for the active invention of tradition and customary practices as people struggle to survive and prosper in increasingly difficult economic and ecological conditions. Competing claims to land in contemporary Africa are frequently based on appeals to contradictory interpretations of history and tradition (see Ranger 1989; Goheen 1992; Berry 1992). Both Scott (1985) and Evans (1987) note that it is the "losers" in the process of changing economic relations who most staunchly defend "tradition" and do so selectively and creatively. Finally, subsistence "guarantees" are sought through attempts to create new relations of mutual obligation or reciprocity that are judged against the standard of a local normative system. The Arusha National Park authority is, within the context of the local economy, in many ways reminiscent of a large estate whose "owner" is seen to have obligations to its "clients." Thus these new relations are sought out at different levels, from deeply personal and intimate ties between villagers and park guards to moralistic appeals to national politicians and officials. Where personal ties and formal appeals fail, various forms of individual and collective resistance may be employed to defend against transgressions of moral economy.

DEBATING PEASANT POLITICS

The African historiography produced in the 1970s (and through the 1980s) was principally concerned with recovering African precolonial history and resistance to colonial rule (Cooper 1994). This shift in focus was complimented by a renewed interest in African peasants and the ways in which rural societies struggled against foreign occupation (Isaacman 1990). While the numerous studies on forms of African peasant resistance produced in the decades of the 1970s and 1980s are thoroughly reviewed elsewhere (ibid.), two conceptual issues warrant discussion here. The first revolves around the very concept of "peasants" as an undifferentiated class. An expanding body of research reveals the limitations of analyzing "peasant resistance" based on a conceptualization of a homogeneous peasantry by demonstrating the multifarious fracturing of African rural societies along lines of class, ethnicity, age, and gender (see, e.g., Carney 1993; Lovett 1994; Schroeder 1993; D. Moore 1993). One of the principle criticisms of analyses of peasant resistance formulated within the moral economy school has been a disregard for the processes of differentiation within agrarian societies. Akram-Lodhi, for instance, charges that Scott's concept of "everyday forms of resistance" (Scott 1985) "is rooted in his essentially neo-populist approach" (Akram-Lodhi 1992, 197) which precludes an analysis of class formation within peasant communities. Regardless of these charges, African peasants inhabit contradictory class positions as petty commodity producers (Bernstein 1988) and may espouse or be subject to contradictory political positions, given their multiple identities in gender, class, religion, and ethnicity (Beinart and Bundy 1987). Generalized statements about "the African peasant" are therefore

untenable, and the political content of the actions and speech of peasants thus understandable only through careful contextualization.

The second conceptual issue I wish to address here is the question of
identifying the actions and speech of peasants as resistance. This is both
a theoretical and methodological question. The theoretical aspect
revolves around the concept of hegemony developed by Antonio Gramsci in *Prison Notebooks* (Gramsci 1971). In contemporary studies of
dominance and resistance, there is a general lack of consensus concerning Gramsci's conceptualization of hegemony, with one camp constru-
ing hegemony to mean ideological quiescence and a lack of independent political consciousness among the peasantry (Arnold 1984; Scott
1985; Feierman 1990) and the other interpreting Gramsci as implying
internalized forms of social control and self-surveillance (Mitchell 1990;
Akram-Lodhi 1992; Lovett 1994). In formulating his ideas of "everyday forms of peasant resistance" Scott traced the lineage of Gramsci's
hegemony to the Marxist notion of "false consciousness" (Scott 1985,
39, 315–17). Such a conceptualization, Scott argues, "ignores the extent
to which most subordinate classes are able . . . to penetrate and demys-
tify prevailing ideology" (317). For Scott, hegemony is absent or incomplete within peasant communities when speech and action challenge the
ideology of dominant social groups. He thus proposes that resistance
can take the form of any act intended to deny or mitigate claims made
on a class by superordinate classes. These "everyday forms of peasant
resistance" are aimed not at reforming the legal order, but at "undoing
its application in practice" (1987, 447). Such divergent actions as "foot
dragging, dissimulation, desertion, false compliance, pilfering, feigned
ignorance, slander, arson, sabotage" (1985) are important "weapons of
the weak" in the day-to-day defense against exploitation.

Scott's representation of Gramsci's hegemony and his conclusions
about the political importance of everyday forms of resistance have been
challenged from several directions. Akram-Lodhi argues that, for Gramsci, hegemony is not something imposed from outside as Scott portrays,
but rather is internalized as a part of the formation of culture. Though
everyday acts of resistance may demonstrate a degree of autonomy
among the underclasses, they do not threaten prevailing power structures and can thus be construed as being contained within the boundaries of hegemony (1992; see also Lovett 1994). Timothy Mitchell has
developed the most challenging critique of Scott. According to
Mitchell, Scott mistakenly confines Gramsci's hegemony to the realm of
ideas and coercion to the physical realm, thus reflecting a deep-seated

split between mind and body prevalent in much contemporary thinking about power. Scott is able to demonstrate the absence of hegemony, Mitchell argues, by relabeling hegemonic effects as "givens" and "obstacles to resistance" and thus defining hegemony so narrowly as to insure its absence. Mitchell counters that when people operate within taken-for-granted limits as to what is politically possible or even conceivable, they are engaged in self-monitoring and have thus internalized the relations of power (1990). Hence, the presence of conflict and resistance does not in itself demonstrate an absence of hegemony.

The methodological aspect in the question of identifying resistance is only slightly less thorny than the theoretical. Whether in historical or contemporary research, uncovering the meaning and intentions of particular acts of resistance, especially illegal acts, calls for a relatively high degree of interpretation. The fundamental difficulty with any study of peasant resistance is that the type of actions under examination are, from the perspective of elites or state officials, by definition "hidden" and "silent." As Feierman explains, "There is good reason for everyday resisters to *avoid* stating their intentions openly if they are to be effective . . . For resistance to be effective, it *must* frustrate the historian" (1990, 42; italics in the original). Any analysis and interpretation, therefore, must revolve around a process of contextualization, akin to Geertz's (1973) "thick description," whereby the meaning of particular actions are "read" within their historical, social, and cultural setting.

A crucial issue in interpreting meaning is gauging the degree of moral legitimacy within peasant communities attributed both to elite or state claims and to violations of the laws that codify those claims. Scott describes "the rich, and historically deep, sub-cultures of resistance" (1986, 29) in peasant communities that give legitimacy to certain criminal acts. Scott argues that by referencing the cultural context (songs, ritual, jokes) of peasant actions, "it should be possible to determine to what degree, and in what ways, peasants actually accept the social order propagated by elites." For example, "[i]f bandits and poachers are made into folkheroes, we can infer that transgressions of elite codes evoke a vicarious admiration" (1985, 41). Similar to themes developed by social historians studying rural crime during transitions to capitalism (see, e.g., Thompson 1975; Hay 1975; Hobsbawm 1985), Scott argues that the "folk culture" of peasants often celebrates the cunning and evasive tactics associated with crimes such as poaching and tax evasion. We can hear in Scott's portrait of subcultures of resistance, echoes of Thompson's description of crowd action in eighteenth-century England: "the crowd were informed by the belief that they were defending traditional rights or customs; and, in general, that they were supported by the wider consensus of the community" (1971, 78). Certain crimes, in the context of locally constituted notions of justice and customary rights, are, in a word, understood to be "legitimate" by members of the underclasses.

In Africa, questions of how to uncover the political content of illegal actions have been addressed in recent studies of peasant resistance. Isaacman (1990) concludes in his review of this literature that multiple interpretations are often possible, with individual intentionality more than likely being "polyvalent" and perceived in different ways by different groups. One of the most relevant efforts in this area is an edited volume—influenced by some of the revisionist social history of rural crime in England cited above—that explores the relationships between crime and resistance and protest in Africa (Crummey 1986). Within the collection, Prochaska's (1986) subtle analysis of the firing of forests in colonial Algeria traces the multiple and shifting meanings conferred on the practice by various groups in society. Seasonal burning in Algeria's highlands was historically an important land management practice, but declared illegal under colonial rule. His study provides an excellent example of how a single act of forest firing is subject to multiple interpretations—as arson, as customary land management, and as political protest. Similarly in Kenya, cattle raiding of non-community members' herds had been a historical practice among the Kalenjin, but was outlawed under the British. When the colonial government outlawed the practice, it succeeded only in altering the social organization of raiding without changing local consciousness concerning the practice (Anderson 1986). Anderson explores the cultural meanings of raiding and the political context of colonial rule to explain why "sixty years of legislation had failed to cultivate a public opinion against stock theft" (142). In sum, both Prochaska's and Anderson's studies reveal how the persistence of customary practices among colonized peoples challenge imposed ideas of rights, criminality, and morality.

CRIME, RESISTANCE, AND NATURE
PRESERVATION ON MOUNT MERU

Studies of crime, protest, and resistance in Africa have uncovered the political character of peasant actions that had heretofore been attributed to "laziness," "ignorance," or "backwardness" by officials of the colonial and postcolonial states. Timothy Mitchell's (1990) critique of everyday forms of resistance notwithstanding, "to ignore the weapons of the weak is to ignore the peasants' principal arsenal" (Isaacman 1990, 33). Collectively, individual acts of resistance often force responses from elite groups or the state, ranging from concessionary actions to violent repression. Mitchell provides an example from his research in Egypt that both demonstrates how everyday acts of resistance force the state to respond and hints at a contradiction in his conceptualization of hegemony. He argues that the transformation of the land rent system eventually required less surveillance as the rural poor internalized the new property relations, yet then describes "a pervasive and everyday policing" (569) of rural life by the state. If the rural underclasses have so thoroughly internalized the new property relations, why is the state responding with increased policing? While Scott may strain too hard to argue the lack of hegemony demonstrated by everyday forms of resistance, Mitchell goes too far in the other direction to discount the political importance of these acts.

Perhaps the principal limitation of the concept of everyday or hidden forms of resistance is its ambiguity, as these actions may embody diverse meanings and intentions. The key to understanding the political meaning and intention of specific actions, particularly criminal acts, revolves around, as Thompson (1971; 1975) demonstrated, an explication of the social and historical context within which they are performed. Especially critical is the character of community response to criminals and criminal acts. In violating property laws, Thompson's eighteenth-century crowds believed they had the wide support of their communities (1971, 78), and the identities of his poachers and vandals in the early-eighteenth-century Hampshire forests were often hidden from the authorities by local villagers (1975). Turning to contemporary Africa, we can contrast the social meaning of two acts of theft to demonstrate the importance of contextualization. The first case involves urban street theft of handbags, packages, and other personal possessions of individuals. In these situations, it is not uncommon for crowds to violently turn on thieves who have been publicly identified by their victims. Tourists are warned to use caution in shouting for help in such a situation as "mob justice" can result in the injury or even death of thieves. Clearly, there is something of a consensus on the immorality of personal theft among the everyday urban crowd. The second case involves a particular incident wherein game scouts entered an African village in pursuit of a poacher. "Crossing the large river which forms the southern boundary of the [game] reserve, the scouts found themselves far outnumbered by an angry mob of villagers" (Adams and McShane 1992, 134). The villagers attacked with stones and two of the scouts drowned as a result. Here the community consensus is clearly on the side of the thief (i.e., the poacher) and the policing authorities are the ones whose actions lack moral legitimacy and who are subject to "mob justice."

My approach in the following chapters is to examine the actions and speech of Meru peasants in the context of the historical development of state conservation and national park laws and their effects on local pro-duction and livelihood strategies. I hope to demonstrate that most of the violations of resource laws in Arusha National Park that conserva-tionists and park authorities attribute to "poverty," "ignorance," and "population pressure" contain a political dimension. While acts such as illegal fuelwood collection, grazing trespass, and forest encroachment may have multiple meanings and intentions, in the context of the park's criminalization of local customary rights of access, they are political insofar as they represent a rejection of the state's claims of ownership and management. I thus seek to (re)politicize national parks and wildlife conservation laws to show how the bounding of scenic, spec-tacular nature in Arusha National Park disrupts production, threatens livelihoods, and thereby transgresses local moral economy. Viewed from the bottom up, national park laws and policies are therefore seen as criminal, and violations of those laws seen as morally justified.

The bulk of the studies of peasant resistance and protest in Africa, particularly those that examine the links with rural crime, have been focused on the work site and on responses to claims on labor and sur-plus. Comparatively little attention has been given to the relevance of state natural resource conservation laws to the character and organiza-tion of peasant politics. Isaacman's (1990) exhaustive review includes studies of everyday resistance surrounding labor, production, and retri-bution, but offers nothing on struggles over forest and natural resource control. More recently, increased attention has been given to the role of national parks in legitimating European conquest, in the formation of racial and national identity, and in structuring rural politics (Ranger 1989; Caruthers 1989; 1994; D. Moore 1993; Neumann 1995a; 1995b). In varying degrees, these studies explore how myths, symbolic representations, and interpretations of history and tradition are critical for legitimating material claims of groups and individuals and are there-fore subject to contestation. In this interest they overlap with a larger body of African research that explores how struggles over land access often revolve around disputes over the appropriate categories and definitions of land and resources, and over competing interpretations of history and tradition (Carney and Watts 1990; Goheen 1992; Berry 1989, 1992; Peters 1995; Ranger 1993; Shipton and Goheen 1992). In sum, these studies demonstrate "how struggles over land and environmental resources are simultaneously struggles over cultural meanings" (D. Moore 1993, 383).

A focus on the interplay between struggles over material access and cultural meanings brings us full circle to the opening exploration of how an Anglo-American landscape aesthetic forms the core of the national park ideal and what that means for rural Africans' land access and production strategies. For most educated Europeans, the landscape of Arusha National Park represents a remnant of "nature" in Africa. For the Meru peasants who herd livestock and cultivate on the slopes of Mount Meru, the landscape embodies their history *in place*, symbolized by ancestral grave sites and sustained by a collectively formulated memory of past occupation and utilization. Under European occupation, land and resources on Mount Meru, once controlled through lineage-based and kinship-based tenure systems, fell under the control of private individuals or the state, consequently restricting access to important household resources. Control of the land and resources on the mountain continued to be a point of struggle between local Meru farmers and the state officials in wildlife and park agencies through the 1980s. No less important, colonial rule introduced a new set of representations and meanings—inseparable from the new property laws and productive activities—regarding land and natural resources. To borrow from Berger (1980, 76) the new meanings embodied in the scenic landscape of Arusha National Park are addressed to the visitor. The pleasing prospects of the park, for most of the Meru peasants living nearby, are incommensurate with their own history and interests in the land.

2

Political and Moral Economy on Mount Meru

Rachel is awake before dawn, rekindling the fire for morning tea in the family's tiny mud-walled kitchen in Nasula Kitongoji (an administrative subdivision of the village). Her adolescent daughter, meanwhile, has brought their sole milk-bearing cow from the kraal (a small wire or brush enclosure) across the compound. Heavy with the child she bears, Rachel emerges from the precariously listing kitchen building and squats beneath the cow. As the first rays of the sun clear the shoulder of Mount Kilimanjaro, she begins to coax the day's half-pint output from the bony animal. By now her teenage son has turned the goats loose from the nighttime protection of the wire kraal. When his mother is finished, he herds the family's cow and two bulls across the village commons and down to the Ngare Nanyuki River, fifteen minutes' walk away, and then heads for school. Rachel readies herself for the day's main task of sowing maize in their field not far from the village center. When she returns late in the afternoon, she will drop off her hoe and set out toward the national park boundary to collect firewood for cooking the evening's meal. It is a particularly bad time to go beating the bush for sticks since the buffalo often doze in the dense, head-high vegetation during the heat of the day. She complains that firewood is getting increasingly difficult to find and the buffalo add an element of terror that makes her reluctant to venture far from the village buildings. Three months earlier, her young nephew had been gored to death by a charging buffalo while tending his family's livestock nearby. The memory of him dying by the road-side while family members rushed to find transport to the hospital two hours away is still vivid.

Unlike most of her neighbors, Rachel was born in this part of Meru, in a place called Ngare Nanyuki. While she was still an in-fant, her parents' home there was bulldozed and burned in a eviction ordered by the British colonial government in 1951. The land where her present house sits and where her family now cultivates and grazes livestock was once owned by one of the settlers who helped to insti-gate the Ngare Nanyuki eviction. She and her husband support them-selves and six children with the crops they raise from 3.2 hectares carved out of the former estate. Most of their money comes from the sale of any surplus maize, millet, and beans harvested from their three plots scattered up to two hours apart. Her husband once held a salaried position as a foreman, but now takes only occasional tempo-rary wage labor. At thirty-eight, Rachel appears stooped with age and at least ten years older. Her seventh child will be born soon. Her life would be easier, she believes, if the government were to allow her and her neighbors to graze livestock and collect fuelwood in the national park.

The geography of Rachel's daily routine—the places she collects firewood, farms, sleeps, and cooks—is largely structured by the history of land alienations and the inflexible segmenting of land uses by suc-cessive colonial administrations. Her poverty, the futility of her culti-vation efforts, and her search for relief in the lands of Arusha National Park can only be fully understood within the context of a struggle for land reaching back to the German colonial period. This chapter exam-ines how colonial rule has shaped the political and moral economies of land on Mount Meru. It demonstrates that politics on Mount Meru were driven by a larger struggle for land between white settlers and Africans in East Africa. In doing so it places Arusha National Park squarely in the center of this historical struggle. The chapter thus be-gins to link peasant resistance to the colonial state's control over land and resources to contemporary conflicts between the park and sur-rounding villages. And it begins to link the livelihood struggles of peasant farmers like Rachel with the nature preservation efforts of the state.

Precolonial Land Use and Settlement

PRECOLONIAL MIGRATIONS
AND CULTURAL IDENTITY

The ancestors of the Meru people migrated to Mount Meru from the Usambara Mountains via Machame on Mount Kilimanjaro. Dating from oral histories recorded during the early colonial period, the Meru arrived about 350 years ago. Luanda (1986, 32) determined that "[w]hen Arusha pioneers recruited the Kidotu age-set on the slopes of Mount Meru in approximately 1821, the Meru already claimed about ten generations." This dates the arrival of the first Meru ancestors to Mount Meru at roughly the early seventeenth century (see also Spear 1996, 216). While the local oral histories that I recorded in 1989 and 1990 and written documents from the colonial period conflict in some of the details,[1] the broad outline of the history of the migration and settlement is consistent. The nucleus of the Waro people (the name the Meru call themselves) was guided to the mountain by the leaders of two clans, the Kaaya and Mbise, which are today the two largest. Along the way the group split. The leader of the Mbise clan, Lamerei, settled high on Mount Meru near the crater floor, while the Kaaya clan settled at Machame on Mount Kilimanjaro. Later, under Juta, the Kaaya clan returned to Lamerei on Meru. Juta's son, Kasarika, became the first in the hereditary line of Meru chiefs (see below) and settled near the Malala River at Akheri Village, where the last and final chief in the line of Kaaya chiefs resides today. The surrounding area in the middle elevations (1,220 to 1,830 meters) on the southern slope is recognized as the core of the Meru homeland and is commonly referred to as "central Meru" or simply "Meru" by English-speaking Meru.

The other Meru clans (around thirty-five today) came from other areas and peoples. The Nnko and Pallanjo clans, for example, originated from the neighboring Maasai; Urio and Nanyaro came from the Chagga at Sanya Juu and Mount Kilimanjaro respectively. An elder described the continuing immigrations: "When the new people arrived they had to see one of the prophets and say they were 'Waro' and not spies who would later bring enemies."[2] Typically the clans were allowed access to uninhabited areas of land, and the patriarchal heads of these localized descent groups would control access to land and water resources within their area. When a pioneer patriarch died, his homestead became a shrine site where a descendent would "approach the earth through the spirits of his ancestors" (Coulson 1971, 199). These sites were generally protected from fire and were off-limits for settlement, cultivation, and grazing. Thus each clan's territorial origins can be traced to where the *tambiko* (shrine ceremony) is performed. The Kaaya clan had sites at Ndoombo and Duluti, Nasary at Sakila, Pallanjo at Kilinga, Urio at Kimaroro (Singisi), Mbise at Njeku (Mount Meru Crater), and Achoo at Urisho (Sura).

For settled agriculturists, the mountain has clear advantages over the surrounding land, and this no doubt was part of its appeal for the early immigrants. As an Mbise clan elder told me, "People came from the low, dry areas because on the mountain they could get lots of honey and food."[3] Perhaps just as important in Meru history, however, is the representation of the mountain as a place of refuge. A Pallanjo clan elder explained that "[o]ther clans have come from Chagga-land, others from Maasai-land because in those times, there were many wars and people were running from the fighting."[4] The theme of refuge as well as the representation of the mountain as the site of the beginning of Meru history emerged repeatedly in the oral histories. This story told by a descendent of Lamerei is typical: "Lamerei Mbise was the first Meru to arrive on the mountain, and he stayed at the site where the two big cedars are now found in the crater. Lamerei had many sons and many wives and many cows. Because his sons were slaughtering his cattle instead of taking care of them, Lamerei moved his cattle up to the crater to keep them safe. At this time the Meru were dispersed because of war, but Lamerei remained high on the mountain, above the fighting."[5] The importance of mountain forests as material and spiritual refuge is a continuing theme in Meru history and should not be overlooked. During the initial phase of German colonial conquest, many Meru fled the troops by scattering into the higher reaches of the forests. Even in the late colonial period, the upper forested slopes provided a sanctuary from conflict. One Kaaya clan elder recalled a time in 1948 when Maasai warriors massed on the northern margins of Meru lands, threatening war and massive cattle raids.[6] As the British administration scrambled to intervene, Meru women and children retreated to the safety of the higher forests.

Elsewhere in Africa, the isolated, relatively inaccessible mountain regions were critical sites of armed resistance to colonial rule (see, e.g.,

Lan 1985; Maloba 1993). On Mount Meru, colonial and independent government authorities eventually reserved the mountains forests as a refuge for the visiting tourist, thereby restricting the access of two Meru clans to their ancestral shrine. After Lamerei's death, the site of his homestead, now located deep within Arusha National Park, became the locale of a yearly ritual for the Mbise and Nnko clans. According to Sori Kibata, "Lamerei met Masika, a young Maasai who was wandering by himself, and invited him to stay until he grew up, and then he gave a daughter for him to marry. Masika was faithful to Lamerei and he would only listen to Masika, and everyone had to go through him. So now the two clans, Mbise and Nnko, go together to the mountain. They take a pair of sheep to leave for Masika in hope that they will multiply, and they expect that he will take them to Lamerei."[7] The mountain is not simply the place where "the Meru" settled but is a major symbolic source of a collective identity. The name "Waro" derives from the Kimeru (a Machame-Chagga dialect) word for ascent. As one elder explained, Waro are the "people who climb." Their history reveals that their emergence as a relatively coherent social group occurred simultaneously with their arrival on Mount Meru. One becomes "Meru" not by sharing an ancient common history but through arriving on the mountain and being accepted into the existing society. It is the mountain and the historic ties that the Meru have to it that help define them as a people; their identities are forged *in place* (see Watts 1988). The Meru place-names and ritual sites on the mountain tell the story of their ancestors[8] and link the present generation to its past through geography. History and place are merged, symbolically and materially; place is thus "the crucible in which *experiences* are contested" (32; italics in the original). The memory of the ancestors' experience on the mountain is contested by the state's declaration of the national park and forest reserve.

I do not intend in this discussion to argue for cultural, political, and socioeconomic homogeneity among the Meru. It seems clear, for example, that in the realm of land and resource access the more crucial identity prior to colonialism was that of family and clan rather than tribal membership. Additionally, there were distinct divisions of labor in regard to the keeping of livestock versus cultivation that fell along clan lines and resulted in significant cultural and economic differentiation within Meru society. The Nnko clan, for example, who trace their ancestry to the Maasai, were noted as cattle keepers, and agreements were made through trade and marriage for Nnko clan members to herd the livestock of other clans. Furthermore, class and religious divisions have multiplied since the initial arrival of Lutheran missionaries in the late nineteenth century. Nevertheless, I want to draw attention to a culturally constituted representation of society-land relationships on Mount Meru that stands in opposition to the rigid separation of land uses that emerged with the introduction of colonial land and resource policy (Beinart 1989). As we have seen, the concept of natural landscapes for contemplation and recreation was defined by the absence of labor and the denial of human history. There was no such landscape dualism in the representation of Meru lands in their oral histories. Quite the contrary, emphasis is placed on the unity of human history and the land. Though the majority of Meru have been Christianized and the earth shrine rituals are rarely performed, the land retains much of its precolonial meaning as a source of identity. After seventy-five years of colonial domination and thirty years of Tanzanian nationalism, Meru cultural identity and ties to the mountain remain strong. As a local teacher explained to me on the night of my arrival on Mount Meru, "You know, the park is all in Meru country. We are in Tanzania, but we are also in Meru."

LAND AND PRODUCTION
ON PRECOLONIAL MOUNT MERU

The Meru had occupied lands as far west as the Themi River until the late nineteenth century, when the Arusha pushed them back (through warfare and assimilation into Arusha society) to their present western limit at Nduruma River.[9] To the east their lands stretched to near the western edge of the Sanya Corridor (about the present boundary for the Arusha and Kilimanjaro regions). In the south, their claims stopped at Mbuguni, and to the north encompassed all of Ngare Nanyuki, reaching near Oldonyo Sambu (see map 3). Much of this northern grazing area was alternately shared and fought over with Maasai pastoralists. An elder related that in the years just preceding the appearance of the Germans, "[t]he Meru met with the Arusha to fight the Maasai, which allowed the Wameru to return to Ngare Nanyuki. The Wameru overcame the Maasai all in one day by attacking them in their *bomas* [homesteads], and the Maasai all retreated to Longido."[10] Other people remember that during the colonial period they grazed their cattle with Maasai cattle in Ngongongare and Ngare Nanyuki and even shared the same *bomas*. In the 1950s,

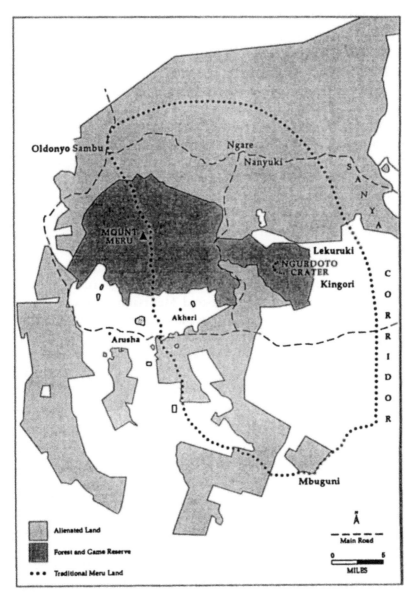

Map 3. Historic Meru land claims (with alienations, ca. 1955). (Tanganyika Territory, Survey Division, Department of Lands and Surveys.)

colonial officials identified mixed Maasai-Meru *bomas* in the Ngare Nanyuki area.[11]

This expansive land claim offered certain opportunities for the organization of production because of the ecological diversity it encompassed. Just prior to European conquest in the late nineteenth century, the Meru were farming bananas, maize, sweet potatoes, beans, and millet and grazing their cattle in fallow fields and in the forests and grasslands of the mountain. Settlement was concentrated on the southern midslope of the mountain in the fertile banana zone between 1,220 to 1,830 meters elevation. Akheri, Poli, Mulala, and Nkoaranga were the principal villages in this area. The grounds immediately surrounding individual houses were interplanted with a wide variety of crops, including numerous varieties of banana, the principal staple. The fertile volcanic soils, augmented by a system of manuring and ditch irrigation, produced ample and continuous harvests. Fields of annual maize crops were planted farther from the homestead. Meru farmers took advantage of the extreme variations in elevation and climate on the mountain by planting crops in different areas during different seasons. Rainfall on Mount Meru is determined by the monsoons in combination with local topography. The large majority of the rainfall occurs in the transitional periods between the monsoons when the Intertropical Convergence Zone (ITCZ) moves across the area, resulting in two seasons of rain: the long rains in late March through May and the short rains in December. The southeasterlies bring the long rains, and therefore the southern slopes get substantially more rainfall, increasing with elevation. To assure a constant supply of agricultural produce and to insure against crop failures, the Meru developed a multiple-plot system and staggered agricultural calendar. For example, maize was planted in highland plots in July and August to be harvested in February and March and in lowland plots in February and March to be harvested in July and August.

The variations in climate and elevation also allowed great seasonal flexibility in livestock grazing. Both the arid lowlands and the highland forests and grasslands were an integral part of the established grazing regime. The northern grasslands and salt licks of Ngare Nanyuki were a critical component of the Meru economy, and cattle owners in central Meru would often have their herd in the care of a relative there. "Milk, butter and meat-on-the-hoof, and animals for ceremonial occasions and compensation purposes in the customary settlement of grievances or for marriage contracts between families, all found their way back from north Meru" (Nelson 1967, 15). They moved their cattle seasonally and the forest and higher elevation grasslands provided grazing for Meru herds, most critically as reserves during the dry season and extended periods of drought. The lowlands down to Mbuguni in the south provided grazing during the wet season for much of Meru.

The basic land designation in their system of permanent agriculture was the *kihamba* (pl. *vihamba*), individually held cultivated plots of land initially cleared from bush and forest. The individual making the initial claim to a piece of land, including land not under cultivation but slated for future expansion, had exclusive rights to cultivate and graze live-stock there. The *kihamba* of the pioneer settler would become the site of the clan's ancestral shrine, maintained at that location by successive generations. With each generation the land under cultivation would be expanded as sons established new households outside those of their fathers. Descendants remained joined to the land and to their families through the maintenance of the ancestral shrines. These patrilineal descent groups, herein referred to as clans, were the principal social unit involved in regulating and managing access to land and resources. Clan elders in council settled land disputes and inheritance claims for their clan. Additionally, land use in and around the shrine sites, as mentioned above, was strictly controlled by the clan concerned. An Mbise elder explained that at the Mbise-Nnko shrine site in Mount Meru Crater, "the whole area was off-limits for any use at all, and the Mbise clan were looking after people keeping hives at Njeku or starting fires. They would fine offenders one cow, two sheep, and *pombe*."[12] Land thus embodied a moral order, which linked the living with their ancestors and social identity with a history forged in place.

Gathering materials in the forest was also an important part of Meru economy. As will become clear in subsequent chapters, honey, building materials, fuelwood, and medicines collected in the forest were crucial elements in household production and reproduction strategies. Hunt-ing wild animals was of lesser importance and, especially among the cattle-keeping clans, was probably looked down upon as a food source. Elephant (*Loxodonta africana*), trapped in large pits, were generally only eaten by elders and the young for ceremonial and medicinal pur-poses respectively. Rhino (*Diceros bicornis*) were for the most part ignored as a meat source. Waterbuck (*Kobus ellipsiprymnnus*) and bush-buck (*Traggelaphus scriptus*) were hunted using rope snares, but were relatively unimportant protein sources compared with livestock, except during famine. Above all, the principal reason for hunting and trapping wild animals appears to have been to clear land for cultivation or prevent crop losses in already cleared areas.

The Colonial Encounter

CONQUEST AND ALIENATION

Land became *the* issue in Meru politics, particularly in regard to the legitimacy of the chieftainship, from the very first experiences with Europeans to the final days of colonialism. Their introduction to Lutheran missionaries was the first close encounter that the Meru had with Europeans in their midst, and that abruptly truncated relationship raised explosive questions about access to land. The Leipzig Lutheran Mission desired to make inroads west from their established center at Machame on Kilimanjaro. Two missionaries, Ewald Ovir and Karl Segebrock, were welcomed to a site on the Malala River in Akheri by Mangi (chief) Matunda in October of 1896. "It was *Mangi* Matunda, in the presence of Captain Johannes, who received the price for the land purchased by Ovir and Segebrock" (Luanda 1986, 31). During their last day on Meru, they "conspicuously fenced in their large plot" (Spear 1994, 117), an action that perhaps sealed their fate. On October 20, a group of Meru warriors, upset over the sale of land to the missionaries, attacked their camp in the predawn hours and killed them both.

Luanda related in detail the subsequent German military retaliation: "On 31 October 1896 Captain Johannes mobilized a massive force consisting of 100 askari and 8,000–10,000 Chagga warriors and proceeded to Meru to punish the Meru and Arusha people. The actual combat took place on 5 November 1896. Meru and Arusha were prepared for the attack and they resisted fiercely. However, after an engagement of three hours in battle, German troops defeated the combatants" (1986, 38–39). After the military defeat, the German commander, Captain Johannes, sent Chagga warriors to loot the Meru villages. Ten thousand head of cattle were taken back to Kilimanjaro from the Arusha and Meru (Luanda 1986). The Meru were, at this point, an impoverished, disarmed, and demoralized people. Luanda relates a personal history of a Chagga farmer who settled in Ndoombo (just above the village of Akheri) after the retaliation: "Ndosi recalled that the country was virtually evacuated—Meru had taken to the forest. He built his hut there. There was plenty of land. Wild animals used to wander around all over the place" (157). The Arusha, the reigning military power on the mountain, sued for peace with the Germans in 1891 and it was granted. By 1900, with the building of the German *boma* in the midst of Arusha country, the military conquest of Mount Meru was complete. In consolidating their conquest of the Meru, the Germans executed two successive chiefs from the hereditary line (see table 1). Lobulu was accused of murder and executed by the Germans in 1900 and replaced by his brother Masenkye, who was himself accused of murder and executed in 1901 after less than a year of rule. Both were from the Kaaya clan in the hereditary line of the chieftainship. After this, the Germans appointed someone from outside of the Kaaya clan, who also lasted only a year. Eventually, they found a chief they could control, Sambekye Nanyaro, from a small and powerless clan originating on Kilimanjaro.

Having cleared the fields and forests of African livestock and driven the Meru into virtual hiding on the mountain, the Germans were free to make claims on "vacant" lands. Alienations took place after the conquest under relatively tight government control, in some cases supervised by Captain Johannes (Luanda 1986). In 1909, "taking the amount of suitable land for crops and livestock given as 195,907 hectares, alienated land represented about 12 percent" (42). There was more land alienation in 1910–12. Sixty percent of the total German alienations was conducted in those two years (Rodemann cited in Luanda 1986). Land was alienated from the northern highlands area, including Ngare Nanyuki, and from a band of land south and just below the main Meru settlements. In addition, the administration designated the mountain slopes above 1,617 meters as both game and forest reserves, essentially curtailing Meru settlement and land uses. Ultimately, the Germans took the best land for large European estates, and the Meru were confined to a fraction of their original territory.

In 1904, Boers, mostly fleeing the Anglo-Boer war in South Africa, began to settle alienated lands in the Ngare Nanyuki area. The arrival of the Boers on Mount Meru did not bode well for wildlife in the area, as they depended upon wild animals for subsistence and ivory and captured wildlife for zoological gardens for cash income. In 1906, Governor Graf von Gotzen remarked that if the Boers did not "abandon their nomadic habits of mainly living on the slaughter of game," it would be best to replace them with German settlers and be "rid of

Table 1 *The Line of Meru Chiefs*

Chief	Clan	Years of Rule
Kisarika	Kaaya	
Malengye	Kaaya	
Samana	Kaaya	
Kyuta	Kaaya	
Rari I	Kaaya	
Sola	Kaaya	
Rari II (Ndemi)	Kaaya	–1887
Matunda	Kaaya	1887–1896
Lobulu	Kaaya	1896–1900
Masengye	Kaaya	1900–1901
Nereu	Nassai	1901–1902
Sambekye	Nanyaro	1902–1922
Sante	Nanyaro	1923–1930
Kishili	Kaaya	1930–1945
Sante	Nanyaro	1945–1952
Sylvanus	Kaaya	1953–1963

(from Puritt 1970, 49).

these game exterminating Boers" (quoted in Luanda 1986, 48). Following the Boers, the first German settlers were impoverished refugees from Russia, who began arriving in 1906. Unskilled as commercial farmers, they remained only a short time and were replaced by Germans from Palestine and the Reich, who were settled in fifty-four farms that lined both sides of the old Arusha-Moshi Road. By 1910 there were eighty-nine farms allocated to German settlers (Luanda 1986)—37,720 hectares in total—which "became almost a European reserve" (Iliffe 1979, 145). "By 1913 the main pattern of European settlement and economy had already been established" (Luanda 1986, 58). On the northern slopes of the mountain and other areas not suited for coffee production, the settlers engaged in cattle raising and dairying. While the alienations did not immediately disrupt African cultivation and settlement, the European farms constituted a de facto privatization of the lower stretches of every major watercourse and river on the mountain, making it difficult or impossible for the Meru to water livestock in customary locations. They also disrupted the established grazing regime on the mountain by greatly inhibiting cattle movement, a situation that became ever more incendiary as Meru livestock numbers recovered from military conquest.

On 1 February 1920 the former colony of German East Africa became the Tanganyika Territory, one of the spoils of war mandated to Britain by the League of Nations. The British kept the land alienations intact, and through the 1920 Proclamation for the Disposal of Enemy Property sold the farms to British settlers. The European settlers in Tanganyika were one of the few groups that seriously imagined Tanganyika as a settler colony on the scale of that in Kenya. By 1937 only 1.31 percent of the total land in the country had been alienated (Iliffe 1979, 303). But this statistic masks the great local impact of alienations in places like Mount Meru, where the best agricultural lands and mildest climate in Tanganyika are found. Spear (1996, 219) notes that the British increased the amount of alienated land around Mount Meru by 81 percent. For the Meru, the impact of colonization was direct and severe—they found themselves hemmed in on all sides by either alienated European estates or government forest and game reserves (see map 4). They were forced to alter existing systems of tenure and land use. *Kihamba* land was now divided among sons, and the amount of land per household in central Meru thus shrank with each successive generation. Agriculture was intensified, especially after the widespread adoption of interplanted coffee in banana groves, and people were forced to send cattle farther afield or bring in fodder for stall-feeding. Tensions over land between households, generations, and clans increased along with population numbers from the 1930s onward. Finally, and most critically, tensions between colonial authorities and Meru peasants over land shaped the politics of resistance to colonialism in Meru.

CREATING AN "AGRICULTURAL SLUM"

The new European estates strung along the south slope of Mount Meru were dedicated mostly to coffee production. The program for European settlement in the northeastern highlands thus created a situation where a white plantation economy operated quite literally side by side with an African peasant economy. The Meru struggled to stay out of wage labor on European farms, and one focus of the struggle was the right to grow coffee. The introduction of coffee into Meru subsistence farming was associated with the Lutheran Missionary effort—a German missionary first introduced it at Nkoaranga Mission—and most of the early Meru growers were converts. The first

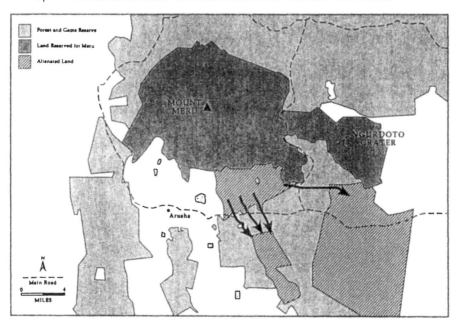

Map 4. Colonial land designations and disputed livestock routes through European farms. (R. P. Neumann.)

Meru to plant coffee did so around 1920 in Akheri without much knowledge about growing or processing the bean. By the early 1920s a handful of Meru growers, mostly Christian converts, were selling small quantities of beans to the settlers in Arusha town and, later in the decade, directly to the Indian and European traders (A. Mbise 1973). The missionary effort thus played a great part in shaping politics and economy. The centers of Christianity were Pole, Nkoaranga, Akheri, and Nkwanrua, which were also the centers of early coffee production and European-organized schools in Meru (see figure 2). Most of the mission schools were built here, which served to reinforce and expand the growing socioeconomic differences as education became a distinguishing characteristic of class among the Meru. As with the Kikuyu in Kenya, it was the mission-educated elite who first articulated African concerns over land scarcity (expressed eventually in the Mau Mau uprising) (see Maloba 1993, 46). This new class of educated, Christianized coffee-growers would later offer the principal political challenge to colonial authority in Meru.

From the beginning, coffee was intercropped with fields of maize,

Figure 2. A missionary school in German East Africa. (Antique postcard, undated, no attribution listed.)

beans, and bananas and grown almost exclusively with household labor. By 1930 there were 374 plots where coffee was grown in Meru. The settlers were not pleased with the Meru's efforts to grow coffee and looked for support from the government to outlaw its cultivation by Africans. Ultimately, the government declined to outlaw African production but in no way could be construed as encouraging the Meru to plant and market it. Thus the sale of coffee, along with the sale of food crops, offered the Meru an alternative to wage labor. The struggle over labor—which Spear (1996) traces to the earliest German contacts on Mount Meru—was part of the evolution of an overall pattern of uneven development in the territory. Iliffe (1979, 151) discusses the differentiation of Tanganyika into labor-exporting and labor-importing regions resulting from the development of a plantation economy. This differentiation had a significant effect on uneven development in the territory, with the labor-exporting zones becoming impoverished and underdeveloped. By 1940, colonialism had produced three economic regions: export crop production on plantations; outlying areas that supplied the plantations; and the labor reserves (Turshen 1984). The Meru, solidly entrenched in peasant production, fell into the second category. Since they were able to refuse to work in the nearby plantations, the government and growers were forced to import labor into the region. This included thousands of landless Kikuyu recruited from Kenya colony to work in government forest reserves on Meru and Kilimanjaro.

With the Meru confined to a small remnant of their historic land claims, increasing population and the growing importance of cash crop production made the land question increasingly critical. If the urgent tone of the district commissioner's correspondence is any indication, by 1930 the politics of land in Meru had reached crisis level. The tone of the officer's prose is agitated; the memo, "Expansion for Wameru Tribe & Cattle," opens excitedly without an introduction: "Apart from certain figures which include the Waarusha and Wameru, (extracts of which are enclosed) I can see no record in this office emphasizing the seriousness of this matter[,] and after 18 months here I have gained some knowledge of the circumstances which I beg to bring before you."[13] The "matter" that he refers to is the burgeoning land crisis in Meru. In his first year and a half there, he had witnessed the expansion of cultivation land at the expense of grazing, "with the result that in certain parts milk is almost unprocurable; the number of milking cows having had to be reduced and the quantity of milk per head being reduced owing to the shortage of feed. This is a most serious matter from a medical point of view, children are therefore deprived of large quantities of milk that they badly need." The establishment of European estates meant that "an iron ring of alienated land was clamped around the native lands."[14] Cattle movements were severely restricted and historic grazing lands were either alienated or cut off from the Meru by European farms (see map 4, p. 64).[15] Stall-feeding was becoming increasingly important, but feed and labor scarcity made it a limited alternative. The end result of the grazing crisis for the Meru was that they had "no other recourse than to resort to the European for grazing and submit to his terms."[16] The agreements that allowed Meru herders to graze on European farms varied but either involved the payment of rent in cash or kind or an exchange of labor. The Meru were not allowed to mix their cattle or even graze on the same pastures as the European owners and so were relegated to the most marginal lands of the estates. The government agricultural officer painted a grim picture of the Meru's predicament: "At present the outlook is uncertain, the tribal areas are rapidly becoming an agricultural slum; alternatives to the people are to become squatters, to try to make a living in the Engare-Nanyuki, Leguruki areas or enter malarious country."[17] He pointed out that many of the European farms were vastly underutilized, claiming, for example, that General Boyd-Moss's and Trappe's farms were stocked with only one head of cattle per 150 hectares. By comparison, in the Meru lands there was one head per approximately .6 hectares. The inequity in grazing land availability is all the more striking given that much of the land in the "Meru Reserve" was inaccessible for cattle, as it was settled or under cultivation.

The foundation of the crisis in grazing access for Meru livestock was a crisis of land scarcity produced by colonial conquest and subsequent alienations of settler estates and forest and game reserves (see table 2). The political and social tensions produced by the land crisis reverberated through all levels of Meru society, from individual households to the institutions of indirect rule. The tensions created over land access would eventually lead to a rejection of the colonial-devised structure of the Meru Native Authority by virtually everyone in Meru excepting the family of the government-appointed *mangi* (chief).

At the root of preindependence political upheaval was the policy of indirect rule, the guiding administrative philosophy in British Tanganyika. When Donald Cameron arrived in Tanganyika in 1925 fresh from an assignment in Nigeria, he brought with him an unquestioned enthusiasm for the ideology of indirect rule (Cameron 1939). The system set up local chiefs and their councils (chosen from clan leaders in Meru) as a legal body called the "Native Authority," established by the 1926 Native Authorities Ordinance. There were Native Authority Courts and Native Authority Treasuries and an assortment of other institutions to take care of implementing colonial law and policy in African society. But by the end of the 1920s, the colonial administration had discarded any enlightened ideals of guiding colonized Africans toward self-governance, and "indirect rule became a means of social control rather than social progress" (Iliffe 1979, 326).

The British authorities, by appointing "chiefs" as local administrators, managed to keep the political struggles localized for a time. But using chiefs and headmen to implement unpopular government policies had the unanticipated effect of conveying authority while simultaneously undermining its legitimacy. Chiefs increasingly came to be seen as the paid employees of the British. This was particularly true in Meru, where the hereditary line of chiefs was broken by the Germans early in the colonial occupation. Eventually the conflicts between the Native Authority and their subjects came to a head in the countryside over the

Table 2 *Complete Game Reserves and Locations as of 1928*

Reserve Name	Administrative District
Katavi Plain[a]	Ufipa
Kilimanjaro[a]	Moshi
Lake Natron[a]	Arusha
Logi Plain[a]	Mpapua
Mount Meru[a]	Arusha
Mtandu River[a]	Kilwa
Mtetesi[a]	Lindi
Ngorongoro Crater	Arusha
Northern Railway[a]	Usambara
Saba River[a]	Dodoma
Selous[b]	Mahenge, Morogoro, and Rufiji
Wami River[a]	Morogoro

[a] Based on game or forest reserves established by the German colonial government
[b] Combined two German game reserves

government's restrictive bylaws for agricultural production. The new regulations covering erosion control, de-stocking, and mandatory crop inspections were much despised, and peasant opposition to them became the most important force behind the growth of the nationalist movement in the rural areas (Bienen 1970; Coulson 1981; cf. Feierman 1990 for an in-depth case study in Shambaai, Tanzania). Because the chiefs were expected to enforce these rules, they were the target of much of the early nationalist politics.

NO DEVELOPMENT WITHOUT TEARS

The obvious impoverishment of the Meru as a result of land alienations prompted various government officers to propose schemes for reallocating land. In the early 1930s, the land development commissioner suggested to the government that some of the forest reserve be offered to the Meru. A few years later the agricultural officer argued that "the time has arrived for the reservation of a large section in the highland forest area for the Meru people."[18] An even more radical proposal was put forth by the local district commissioner to confiscate underutilized European estates and give them to the Meru.[19] As it turned out, the government addressed the problem by doing quite the opposite. A government white paper published in 1952 provides a concise history of the administration's response to the crisis:

The problem of providing additional land for both Arusha and Meru peoples was investigated by a Special Commissioner in 1929/30[,] following which two further farms were acquired by purchase by the Meru ... In 1939 the Central Development Committee recommended for further investigation a proposal that the areas of land between the Kingori and Sanya Rivers be given to the Meru in exchange for the major portion of the so-called "North Meru Reserve" and areas occupied by these tribesmen in the middle of non-native settled areas. War, however, intervened and no action was possible for some years. In 1944 the Post-war Planning Committee recommended that after the war an authoritative commission should be appointed to formulate a comprehensive plan for the redistribution of alienated and tribal land on and around Kilimanjaro and Meru mountains.[20]

Subsequently the government commissioned a study of the land problems in the northeastern highlands in 1946 under Judge Mark Wilson. The terms of reference for the resulting Arusha-Moshi Lands Commission (A-MLC) were to improve the homogeneity of settlement for both Africans and settlers, to relieve congestion in African lands, especially in regard to access to adequate grazing lands, and to look into the availability of lands for future alienations for Europeans. The recommendations that followed resulted in the forced eviction of three thousand Meru and their livestock from their lands in Ngare Nanyuki on 17 November 1951.

The Meru Land Case is a relatively well-documented incident in the chronicles of the East African independence movement and has been documented by some of the principals involved and others (I. Mbise 1974; Japhet and Seaton 1967; Nelson 1967; Luanda 1986). There is no need to cover the same ground in detail here. It is, however, crucial to reexamine the case as the single most important event in the history of Meru resistance to the loss of customary claims on land and natural resources. As such, it left an indelible imprint on the political consciousness of the Meru. The eviction of Africans to make way for European settlers also ultimately served as a catalyst for the development of nationalist politics in Tanganyika. Thus it was part of the larger struggles over land that were the foundation of resistance to colonialism in East Africa, exemplified by the Mau Mau uprising in Kenya. The Meru Land Case, furthermore, is of central importance in understanding the ways in which the events of the colonial era help structure local forms of resistance and protest to the continuing crisis of land and resource scarcity. As my analysis of the politics of nature preservation proceeds in later chapters, I will show how the symbolic capital embodied in the story of the eviction and the subsequent resistance movement is reinvested in continuing struggles over land between the Meru and the state.

The main spark that set the eviction and subsequent events in motion was the idea of a group of settlers, led by M. S. duToit, W. T. Malan, and W. R. Jacobs, to create another "White Highlands" stretching from Mount Meru to Mount Kilimanjaro.[21] Hence the mandate "to improve the homogeneity of settlement" in the A-MLC's terms of reference. At the time the committee's report (known as the Wilson Report) was released in 1947, there were three thousand Meru living on farm 31 in Ngare Nanyuki and farm 328 in Leguruki. These farms were originally alienated by the Germans, then sold back to the Meru by the British and were now surrounded by European settlers. The Meru would have to move from them if racial homogeneity was to be achieved. The Meru were to be "given" land at Kingori in exchange for the land at Ngare Nanyuki. The Meru claimed both areas as part of their traditional lands and questioned how it was they were offered in exchange what they already owned. Though part of their land claims, Kingori was not a desirable place to live because it was infested with tsetse fly and mosquitoes and its chief function was that of a seasonal grazing reserve. In response to these protests, the Northern Province commissioner observed that "development will not take place without tears."[22] The colonial government contested the claim by the Meru that Ngare Nanyuki was their land and that it had been critical to their grazing regime for many generations. The administration considered it vacant land used only by an occasional group of nomadic Maasai (Japhet and Seaton 1967) and that therefore the Meru had no moral ground on which to stand. Besides, Africans "should understand that the loss of land . . . is an inevitable part of the price [they] must pay for [their] advancement."[23]

When the enabling bill for the relocation came before the Tanganyika Legislative Council, the Meru, by unfortunate coincidence or devious legislative scheduling, were left unrepresented. The term of council member Abdiel Shangali, who had been fully briefed by the Meru chief to oppose the bill, had expired just prior to the vote.[24] Once the bill was approved, the government began pressuring the Meru Native Authority to accept the relocation plan. Mangi Sante caved into the government's threats to withhold his salary and accepted the eviction order, in effect signing away what little support he still claimed among the Meru. As Mangi Sante's shrinking reservoir of legitimacy evaporated completely, an alternative to the Native Authority emerged with the aid and cooperation of the Kilimanjaro Citizens Union (KCU). Initially known as the Kilimanjaro-Meru Citizens Union, the Meru Citizens Union (MCU) split off on its own on 1 January 1951 (Japhet and Seaton 1967, 18), and became the major source of organized resistance to colonial authority in Meru. The political strength of the MCU was such that it was able to garner almost universal support among the Meru, with the exception of Mangi Sante and his small (mostly family) following. Strong class, religious, and gender divisions were transcended in opposition to British relocation plans. In London, Colonial Office officials found it "rather a shock" to discover that not just those affected by the relocation, but people from all over Meru territory, were mobilizing in protest.[25] Ultimately, the MCU—in combination with the Lutheran church (at least two organizing members of MCU were Lutheran evangelists)—organized a program of noncooperation and passive resistance to the proposed eviction.

The Meru refused to give the names of the families living in the area or to participate in itemizing the families' possessions. In 1951, Mzee (respectful title for elders) Abraham, a former Lutheran evangelist, was serving as a *jumbe* (headman) in the area. After months of stalling on preparations for the relocation and repeated written appeals to colonial authorities, he was approached by the provincial commissioner (P.C.). Abraham recalled that day, explaining, "[I]f someone buys land, you can't claim it back . . . So I told them, I don't want to work [as headman] anymore. They said, '*Jumbe,* write down the farms, the houses, how many rooms in each house and what belongs to whom; everything.' I didn't agree and told the P.C. and Mangi Sante that I will never write this down and I am no longer a *jumbe.*"[26] Upon resigning his post, Abraham was labeled a "political agitator" and found it necessary to go underground using an assumed name and, ironically, hiding on the estate of a European.[27]

When it came time to move, the residents of Ngare Nanyuki stood to the side and refused to participate in the moving and transport of their belongings. In retelling the story before the United Nations Trusteeship Council, Kirilo Japhet recalled, "We held our own meeting on 17 November, and decided that we would bear anything that was done to us by the Tanganyika Government. We said that we Meru were a small tribe; we had no rifles; we had no bombs; we just had to wait and see what would be done to us" (quoted in Japhet and Seaton 1967, 37). He went on to claim that the government burned the houses to the ground with all the food and possessions inside, including small livestock. He testified that one evicted pregnant woman gave birth in the bush and her baby died four days later, while seven other women suffered miscarriages. Decades later, a Kaaya clan elder remembered watching from the vantage point of Kilima Mbuzi (a small hill in Ngare Nanyuki) with his brother-in-law as the British burned the settlement.[28] He and his wife recalled the ordeal of hungry children who had been turned out of their homes. More in disbelief than in bitterness, they described the brutal treatment that Africans received following the eviction if they were caught in the new "whites only" area.

Kirilo Japhet eventually represented the Meru in an appearance before the United Nations Trusteeship Council, and their appeals ultimately reached the General Assembly, where they failed to get the needed votes to reverse the eviction. That the Meru ended up pleading their case before the United Nations Trusteeship Council is testimony to their own organizing skills as well as a reflection of the importance of land in the emerging nationalist politics of East Africa. Almost from the beginning, the tiny resistance movement in Meru was part of the growing political turmoil surrounding African land rights in Kenya and Tanganyika. Months before the eviction, members of the Meru Citizens Union journeyed to Nairobi to seek the advice of the Kenya African Union (KAU), which was embroiled in a much more extensive and violent land conflict of its own, called Mau Mau by the British government. Abraham, one of the Meru delegates, remembers their involvement with KAU this way: "We had one man living at the cattle dip area at Olkangwado who was a Kikuyu. His father came to visit him. He heard about the plan to move people out of Ngare Nanyuki. He came to visit us at Kilima Mbuzi. He told us, nothing or no one will take your land because it is God who gave you this land, and it is God who gave you this hill. He took a handful of soil, put it into his mouth, and said let us go, to Kenya, I will lead your delegation to Jomo Kenyatta. Jomo Kenyatta gave us the addresses for the United Nations members."[29] This encounter between Meru leaders and Kikuyu immigrants was not particularly unusual. Displaced Kikuyu farmers had been coming to the Northern Province of Tanganyika since 1928, when they were brought into the Kilimanjaro Forest Reserve as squatter-laborers under the Forest Department's *taungya* system, whereby squatters are given temporary cultivation plots in the forest reserve in exchange for their labor on Forest Department projects.[30] By 1952 there were between twelve and fifteen thousand Kikuyu in the Northern Province. Kenyatta had visited these communities three times in 1951 alone.[31]

As land conflicts intensified, the colonial administration and white settlers were fearful that Mau Mau in Kenya would spread to northern Tanganyika through a Kikuyu network. In the aftermath of their compulsory eviction from Ngare Nanyuki, authorities became increasingly concerned that the Meru were "susceptible" to Mau Mau.[32] British intelligence sources knew that the KAU was maintaining close ties to the Kilimanjaro Union and that former Chief Koinage from Kenya (an early organizer of KAU) had addressed a meeting of Meru, Arusha, and Maasai activists in Tanganyika.[33] Desiring to keep Mau Mau out of Tanganyika and believing that "these Kikuyu are too close for comfort to the Wameru,"[34] the administration began mass deportations of Kikuyu in 1952. Though there was no sign that the Meru had "yet been infected with the Mau Mau doctrine,"[35] Governor Edward Twining noted a "general state of unrest . . . and a truculent attitude by the dissident Meru leaders."[36] In the years immediately following the eviction, arson, cattle thefts, violence, and threats of violence, all directed at European settlers, escalated. Meru were suspected of setting fire to European grazing lands on estates they had once occupied, and there were numerous complaints from settlers of trespass and cattle thefts.[37] As some Meru threatened to forcefully reoccupy their lost lands, Twining, seeing the Kikuyu's influence in the disturbances, "decided to smash the militant Mau Mau organizations in [the] Northern Province" in a series of surprise raids.[38]

The land crisis on Mount Meru thus produced a political climate in which agitation, protest, and resistance were the norm in the years following World War II. The government-appointed authority, Mangi Sante, was allegedly the target of repeated death threats from his political opponents. In April 1949, eleven Meru "dissidents" were tried and deported for challenging Sante's rule.[39] In 1948, the Arusha district commissioner believed that a state of emergency existed in parts of Meru, and in 1951 the provincial commissioner threatened to declare the area a "disturbed area" and post police at tribal expense.[40] The antieviction movement was, in essence, part of a larger anti–Native Authority movement and, more generally, a challenge to the legitimacy of all colonial land alienations. The Lutheran church, abandoned by the

Germans and now led by German-trained African pastors, essentially supplanted a weak and delegitimized Mangi and Native Authority as the focus of political organization.

Christianized peasant coffee-growers had been gradually gaining control of politics on the mountain, culminating in the resistance to the Ngare Nanyuki eviction. The mission-educated leaders of the Meru resistance movement shifted in and out of low-level clerical positions with the government, church, and settler farms. Abraham's life history epitomizes the new elite. Educated at a Lutheran Mission in central Meru, Abraham helped build a church near Ngare Nanyuki in 1929. He recalled, "In 1932, I came to Ngare Nanyuki, Kilima Mbuzi, to teach bush school and conduct church services . . . In 1937, I returned to Mulala to get married. In 1938, I returned to Ngare Nanyuki and stayed at Monas's farm. I worked as a clerk for the farm and was a foreman for the laborers . . . Then I was called to work for the church again . . . In 1947 I was called by the people of Ngare Nanyuki to take the post of *jumbe* in the British administration at that time."[41] Abraham held the *jumbe* position until he resigned in protest over the eviction and fled to the sanctuary of a European estate.

The colonial clerks and mission evangelists like Abraham were the people Feierman labeled "peasant intellectuals," the local leaders "who elaborated the discourse of the movements of peasant resistance" (1990, 23). Though a minority among the mostly "pagan" Meru, they were able to organize a fractious and divided society into unified resistance to European occupation. Their political strength was greatly enhanced during the eviction by their ability to produce and maintain nearly complete solidarity not only in Ngare Nanyuki but throughout Meru. The MCU and the associated church leaders extended their organizing skills to other land struggles on the mountain. In the years following the eviction, two freehold properties below the southern Meru Reserve, farms 90 and 91, became a second focus of organized resistance. The Christianized coffee growers lent their support to a group of squatters who were refusing to pay rent or vacate the properties, owned by a settler named J. Focsaner. About thirty Meru families lived there (some had lived there for up to thirty years), "performing annual labour for the owner and paying a peppercorn rent. Some three or four years ago these people refused to perform further labour, to pay rent or to vacate the land."[42] On 10 March 1952 the Arusha district commissioner attended a meeting of about 150 Meru peasants protesting the closure of a cattle track through the same farm.[43] Despite the ongoing political unrest over land on Mount Meru, the commissioner ordered the track closed and the resident magistrate ordered the squatters to vacate the estate. The Meru Native Authority police, however, refused to carry out the order, and the families remained squatting illegally. The government merely inflamed the already incendiary relations between settlers and the Meru. Settlers soon reported "the frequent cutting of the wire fence which forms the Meru Corridor in this area, and the consequent large increase in trespass."[44] Focsaner, the settler who owned the farm through which the track passed, complained about Meru "trespassers" threatening his non-Meru workers with *pangas* (machetes).[45]

For the Meru, the critical issue remained the land crisis resulting from European alienations. While MCU members espoused the illegitimacy of European land claims, Governor Twining assured the Meru: "Those that state that there is no such thing as alienated land and that all forests and rivers and the leasehold and freehold land belong to the tribe are giving a further display of their ignorance."[46] The fallout from the Meru eviction and other land struggles would eventually spread beyond the northern highlands to fuel the growth of the nationalist party, the Tanganyika African Association, later Tanganyika African National Union (TANU).[47] The critical aspect of the Meru Land Case was that it carried African grievances beyond the local to question the very legitimacy of colonial rule. On Meru, it also served to thoroughly discredit Mangi Sante and the Meru Native Authority. Consequently Sante resigned in 1952, a new constitution was drawn up, and Sylvanus Kaaya was voted in as the new *mangi* in a general election. Part of the hereditary line of Meru chiefs, Sylvanus was a third-generation Lutheran, a coffee grower, and a former medical assistant in the government dispensary at Ngare Nanyuki.

The various local and official colonial narratives of the land crisis in Meru and the responses of Meru peasants are important to relate here for a number of reasons. The Meru Land Case has become semi-mythologized in local culture, representing the gross injustices of colonial rule and the heroic resistance of a generation of Meru leaders. Personal accounts of the eviction and its aftermath are told freely to outsiders, often without solicitation and always with a certitude of the righteousness of Meru land claims. The case is illustrative of the way in which a society sharply divided by religion, class, age, and lineage can present a unified front of resistance in the face of onerous outside claims (see, e.g., Peluso 1992; Thompson 1975; Isaacman 1980). Even the agents of indirect rule, in the personage of the Meru Native Authority police and headmen, participated in the resistance to the loss of land and resource access. The historical narratives also place the Meru in the thick of the nationalist movements and anticolonial rebellions that erupted in East Africa after World War II. The support for nationalist politics among East African peasantries was largely based on agitation over the burgeoning land crises of the 1940s and 1950s. Interestingly, as will be investigated in a subsequent chapter, the politics of land and nationalism also provided the historical context for the first national park in Tanzania, Serengeti, and many of the same players in Mau Mau and the KCU and MCU appear in that struggle.

The Meru Land Case is locally of great symbolic importance. Colonialism has today become a metaphor for the perceived injustices perpetrated by outsiders, including government officials. As we will see, *kama ukoloni* (like colonialism) is a phrase used in local critiques of Arusha National Park policies. The reference to colonialism, however, is not merely metaphorical. The lands of the national park are either former settler estates or colonial forest reserve, all of which had been alienated by the close of German rule. The park thus serves as a persistent reminder of the continuing social, economic, and moral costs of past land alienations.

It also serves to remind that the support that peasantry gave to the nationalists ultimately was not rewarded by land reform when TANU took over Tanganyika from the British. After independence, TANU became the power base of a ruling class of bureaucratic bourgeoisie who, though having no economic base, seized control of the state apparatus (Shivji 1976). The end result was an economic bureaucracy serving as industry managers, whose roots lay in the international bourgeoisie and who had an economic interest in maintaining the status quo, rather than an interest in addressing the political and economic concerns of the peasantry (Bolton 1985). The position of the peasantry in this postcolonial political economy is summed up concisely by Feierman: "In the 1950s the peasantry fought to limit the power of the chiefs. They found allies in the nationalist party. This left an independent Tanzania in the 1960s with only two main centers of gravity: peasants and bureaucrats, with the latter working to establish their sole dominance . . . The new government continued and elaborated a policy of bureaucratic control over the day-to-day workings of peasant agriculture . . . Bureaucratic control over the peasantry, which had been called *indirect rule* under the British and the chiefs, now continued within the context of African socialism" (1990, 22–23). The colonial institutions, such as the Divisions of Forestry and Wildlife and Tanzania National Parks (TANAPA) remain today, run now by a class of educated African bureaucrats rather than Europeans. For Meru peasants squeezed by the forest reserve and alienated European estates, independence brought little improvement in terms of local control over land and natural resource access.

Land, Production, and Settlement in Contemporary Meru

Much of the colonial pattern of land distribution on Mount Meru has remained unchanged since independence. The only en masse evacuations of Europeans at independence were those of the Boers living on the relatively marginal agricultural lands in the north. The most fertile lands have remained as coffee estates in the hands of individuals or as parastatal operations. Land scarcity remains a critical problem and land availability a moral and political issue. When then-President Ali Hassan Mwinyi toured the area in early 1990, he told people that the government could not redistribute land and that they should move to another area if there was a land shortage. "If we seize the estates, it would mean that the government has no stand," Mwinyi told a public gathering in Patandi Village on Mount Meru. The continuing tensions over land access, felt throughout Meru from the level of the household to the national political arena, are the principal colonial inheritance.

LAND AND SETTLEMENT IN CENTRAL MERU

Today the historical center of Meru settlement is one of the most densely populated rural areas in East Africa. Tanzania's 1988 national census (United Republic of Tanzania 1989) recorded 14,038 people living in the ward of Akheri (in the heart of central Meru), an increase of 17.4 percent from the 1978 total of 11,597 (United Republic of Tanzania 1981). Since 1921, population density in Akheri has increased from 44 to 304 persons per square kilometer in 1988.[48] Electricity reaches most of the areas where coffee production provides enough cash to pay for the service. Houses built of concrete and wood are common and increasingly replacing mud-and-branch structures. Over the decades, population growth, in combination with the alienation and concentration of land into large holdings, has produced a complex variety of responses in livelihood strategies. Specifically, people have responded by intensifying agricultural production, increasing their involvement in external labor markets, and migrating to less densely settled areas on the mountain.

The intensification of production entailed changes in the management and spatial arrangement of crop and livestock production. From a distance of a few kilometers, the slopes of Mount Meru appear covered by thick, verdant forests. It is indeed thickly covered in trees, but it is also an intensively managed environment. As noted above, the intercropping of coffee with food crops, manuring, and irrigation of the plots immediately surrounding household compounds facilitates increased food production while simultaneously allowing production for the world market. Thus, excepting the steepest slopes, central Meru is densely intercropped with a variety of food staples, principally bananas, beans, maize, squash, papaya, millet, and potato cultivars, in combination with the main cash crop, coffee (see figure 3). Those households controlling fields beyond their immediate compound are able to cultivate separate plots of maize and beans. In recent years, maize and bean fields have progressively been expanded up the previously uncultivated steep slopes of narrow river valleys. Various species of fuel and timber trees are planted throughout all types of cultivation plots, and often as boundaries between holdings.

Livestock keeping has also intensified. As a result of the historic land alienations and steady population growth, lands in central Meru once used as cattle pasture have been converted to cultivated land. One solution has been to keep livestock, including cattle, sheep, goats, and donkeys, in the northern highlands or the southern plains, often tended by extended family members on common pastures. This is a modification of the historical movement of cattle between central Meru and the pasture lands to the north and south. The other solution is to keep fewer cattle, with greater milk-producing capacity, and stall-feed rather than pasture livestock. The keeping of improved breeds of cattle, such as Jerseys and Frisians, is common in Akheri. The practice requires, however, a large initial cash investment and recurring cash expenditures for expensive vaccinations and veterinary treatment that is beyond the means of many farmers, particularly the young. Crop residues from household plots, especially banana leaves and maize stalks, are brought

Figure 3. A densely intercropped field of coffee, banana, and various food crops in the family compound in Akheri. (R. P. Neumann.)

to stalls within the family compounds. This is supplemented with grasses cut from the large coffee estates at the base of the mountain, hauled up to Akheri on the heads of wives and daughters or by hired boys who pile the crops on donkey carts and bicycles.

Despite the successful efforts to intensify production, smallholders cannot survive on farming and livestock keeping alone. Typically, one or more members of a household work for a wage in Arusha, the country's second largest city, only a thirty-minute walk down the mountain and twenty-minute bus ride away. Elijah's situation is illustrative. Elijah has exclusive control of only the .2-hectare plot that surrounds his house. He keeps two Frisian cows in the stall next to his house, where his wife and four of five children live. The plot is densely intercropped with six varieties of banana, potatoes, beans, assorted vegetables, and coffee. He has in past years also negotiated access to portions of his father's extensive landholdings elsewhere. Some planting seasons he is unable to take advantage of this access because he lacks the needed capital to buy seed, hire plowing, and purchase other inputs. In addition to farming, Elijah holds a full-time salaried position at a factory in Arusha town. His wife also works full-time as a nurse in a local hospital. He would happily quit his position in town and devote all of his time to farming if he could devise a way to make it financially viable. But he finds himself in a dilemma common in Akheri. He has neither enough land nor capital to make a living as a farmer, and the couple's combined salaries are not sufficient to allow them to abandon farming altogether.

Another aspect of the land crisis on Meru is increasing landlessness, particularly for the younger generation. As noted above, one of the principal changes in the land tenure system brought about by land alienation was the division of a father's *kihamba* land among sons. The inheritance land handed down from fathers to sons falls within the jurisdiction of customary law, adjudicated by senior male clan members. Clan elders meet in council weekly to arbitrate in matters of marriage and bride price, compensation for injuries and death, and land disputes and inheritance. The land decisions of the council have grown ever more contentious as population density increases. With each successive generation, the inherited plots of sons have grown smaller, increasing tensions between fathers and sons, elders and youths, over the allocation of land. Both the young men and the elders accuse one another of "not following traditions." The elders say the youth do not show respect, and young men say the elders are not meeting their obligations to their sons.

The tensions between elders and youths over the meaning of traditional practices is one aspect of the struggle within Akheri to maintain the moral economy, specifically the right of all to have access to land. The land crisis on Meru is an ongoing "moral crisis" (Spear 1996, 234) as much as a crisis of production and reproduction. The growing land scarcity and the contradictions between the ideal land tenure and reality is straining Akheri's social fabric. As part of the moral economy of land, customary *kihamba* land is ideally considered not fully alienable. That is, customary land should not be transferred out of the immediate family, or, in the absence of a direct heir, should remain in the clan. In reality, there has been a land market for some years, with plots transferred to local creditors when loans fall in arrears (see Moore and Puritt 1977). Nevertheless, the unalienability of *kihamba* continues to be the ideal, and when customary land falls into an outsider's possession every effort is made to bring it back into the family at the earliest opportunity. When *kihamba* land does fall into the hands of non-family members, fathers are accused by sons of squandering their inheritance. Fearing the loss of their main means of acquiring land, sons put great pressure on fathers to transfer rights sooner. For example, a farmer and entrepreneur in Akheri boasted to me that he threatened his father with an ax in demanding his "rightful" land allocation.[49] Regardless of whether the encounter actually occurred, the story demonstrates the frustration of young men and the level of intergenerational tensions over land.

The crisis in the moral economy of land sends reverberations throughout most aspects of social life in Akheri. For example, the meaning of marriage and the mutual obligations of husbands and wives is changing. Land scarcity has undermined a central pillar of the marriage contract, the husbands' obligation to provide their wives with their own plots for cultivation. As a result, more young women eschew marriage, seeing little economic advantage and mostly increasing labor burdens. Since women generally do not inherit land and there is virtually no land for sale, their prospects are limited. Young women with less than a university education often exist in a sort of economic limbo. Not skilled enough to command a wage to support an independent life, yet lacking opportunities in the village, they take service jobs in town but remain in their fathers' households. The low wages they earn barely cover the costs of commuting, clothing, and food, a situation that provides almost no chance for economic improvement. Women, therefore, find themselves in a socioeconomic position more precarious than that of young men.

Inevitably, some sons, usually the middle sons, are left with no land inheritance at all. The choices, in this case, are limited: stay and attempt to live exclusively on wage earnings or move out of the village. Some have taken land in other districts or even other regions as part of government land allocation programs. Most have chosen to move to other parts of Meru, to the west in villages such as Maji ya Chai or Ngurdoto, or further north to Olkangwado. The expansion of settlement into these villages is complex, because the pattern of migration has been so structured by the spatial and political characteristics of colonial land alienations. Some of the areas of recent colonizations were part of the lands "reserved" by the British government for the Meru, and other areas were part of former European estates. Two of these villages, Olkangwado and Ngurdoto, now share boundaries with Arusha National Park.

LAND AND SETTLEMENT IN NASULA

Nasula Kitongoji (subvillage) lies on the northern boundary of Arusha National Park and is a subdivision of Olkangwado Kijiji (village). Tanzania's 1988 national census (United Republic of

Tanzania, 1989) recorded 3,643 people living in Olkangwado, an increase of 21.4 percent from the 1978 total of 2,864 (United Republic of Tanzania 1981). The *kitongoji* was settled by people migrating from other parts of Meru—including women who have married into the community—and their offspring are now establishing households of their own here. Mostly they came from the congested villages of central Meru, Songoro, Sura, Urisho, Mulala, and Nkoaranga "to look for more land."[50] The origin of Nasula Kitongoji is closely associated with the eviction from Engare Nanyuki and the later flight of European settlers from the northern highlands immediately following independence. Located about 5.5 kilometers north of Nasula's cluster of houses, Olkangwado Village's administrative center and market occupies the site where the Tanganyikan colonial police burned the homes and fields of the evicted families.

Most of Nasula's lands were formerly a freehold estate owned by a settler named A. S. Monas. A small but socially and politically important portion of the village is comprised of a corner of the Trappe family's former Momella Farm.

The history of the alienation, dislocation, and reoccupation and the creation of Arusha National Park have combined to shape the spatial pattern of settlement and land use in Nasula. The beginnings of settlement can be traced to a small group of Meru squatters who found employment and grazing access on Trappe's estate, after being driven from Ngare Nanyuki. Mzee Molel, probably approaching ninety years of age at the time of this study, was one of those squatters. Molel originally came to Olkangwado from central Meru "to graze cattle" during the reign of Sambekye (1902–22).[51] He was one of the group that moved there after the British government sold the farm originally alienated by the Germans back to the Meru. The reacquired farm presented new pasture access and an opportunity for young men like Molel to build their cattle herds and avoid working for Europeans. "When we moved to this place, I was not a worker," he explained; "that was after Ngare Nanyuki was set fire to by the British government."

Evictees like Molel had to scramble to find relatives or friends with whom they could stay and new pastures where they could graze their livestock. After the eviction, the government offered monetary compensation and resettlement funds in the Kingori area, but only Mangi Sante and his immediate family accepted the deal. Most Meru continued to protest in solidarity by refusing any government assistance or compensation. Molel managed to find refuge on Trappe's freehold estate. Trappe, a German who had been on Mount Meru since 1907 (excluding several years of exile after the mass enemy deportations following World War I), and Molel had become acquainted over the years. "[After the eviction] there was a certain German [Trappe] and we knew each other. This man sent a message to me, saying that I should come to work in his farm and then to get a piece of land there." Molel and his friend Simera remember the Trappes as patrons. Believing that Trappe was sympathetic to their cause because of his own deportation by the British, Simera recalled: "Trappe's wife managed to collect the maize for Molel in his burnt homestead. Then she kept the maize and asked Trappe to send a message so that Molel could come to their farm. So Trappe called Molel back. Molel came to Trappe's farm later on with his relatives . . . This allowed them later on the chance of acquiring [access to] the Greek's [Monas's] farm." Molel came and settled in 1952 in the area where Momella Lodge now stands. He brought with him some of the fugitives who were wanted for their role in organizing the resistance to the relocation.

Molel and the others provided labor to the Trappes' enterprises in exchange for a plot to cultivate and for pasture for their cattle. They lived and worked there, but grazed the bulk of their herds on A. S. Monas's estate just to the north, sometimes for rent in kind, sometimes in cash. "We paid a lot of money because we had a lot of cows to graze," Molel remembered. By 1958, Trappe and his wife had passed away, leaving their son Rolf an insolvent estate. To raise capital, he began selling off land. In 1961, Trappe sold about four hundred hectares of his freehold, including the area occupied by the squatters, to Hardy Kruger, a film actor who had worked on the John Wayne film *Hatari* (filmed on location at Momella), and James Mallory, an investor interested in lodge development. After the lodge was built and Mallory took over as managing director, the squatters "continued grazing as before." However, they moved their houses, because, according to Molel, they "didn't want to live next to *wageni*" (literally guests, but often used to refer to tourists). They built new homes in 1964 on a piece of the former Monas estate, just across the old property line shared with Trappe.

They continued to use the Trappe area surrounding the lodge as a grazing commons, and years later built the Nasula primary school on the land between their new homes and the lodge. By the time they moved their houses, Monas had died and most of his Boer neighbors had sold their farms back to the Meru and left the northern highlands. Apparently, Monas's surviving sons allowed the squatters and a farm employee to take over his estate lands without seeking compensation. Two slightly different versions of the transfer are recalled by the Trappe squatters and a former employee of Monas. Molel recalls that when Tanganyika won independence, the "government calculated that the total payment of money which we had been paying to the Greek for the whole period was equal to compensation for the farm . . . [Monas] said this farm had already been bought by certain people for each year they paid rent." Mzee Joseph, a longtime friend to Molel and an employee of Monas on his farm during the 1950s, remembers that Monas gave the farm to him for faithful service.

Though his father was not originally of the Meru, Joseph was born in Ngare Nanyuki and then moved with his father's family to Leguruki, after the British destroyed their home in the 1951 eviction.[52] Sometime in the mid-1950s he returned to the area to work on Monas's estate. Monas had large landholdings throughout Tanganyika, including a sisal estate, and had left the holding on Mount Meru in the charge of a European estate manager named Paul. Joseph worked as his assistant, serving as foreman for hired African labor. During an extended period in the late 1950s when the estate manager was sick and mostly bedridden, Joseph nursed him and took care of the day-to-day running of the farm: "Then I came to stay with Paul in 1959, and he was seriously ill. He was so sick, he gave me money for making his coffin . . . Paul had relatives living at Machame-Moshi area: a brother-in-law named Mr. Vasil. I had sent a message to him and he came to see Paul. Paul was taken away by his in-laws and the same day died. But he had written a letter in Greek and left it to me." Joseph was unable to read the letter, but when Monas's wife and son came sometime later to settle affairs, they told him that the letter contained the details of how Joseph had cared for Paul and stopped a neighboring settler from stealing the payroll while Paul was bedridden. "Monas's wife and son checked the message . . . and said, . . . 'Paul wrote that you were like his son and directed that you take charge of the farm and everything.' So Monas's wife and son handed everything with the farm to me because I had worked here for so long. It was like a pension." The Monas estate was a freehold, but the actual title transfer apparently was never recorded; Joseph has only the letter in Greek. With the coming of Tanganyika's independence, Monas's wife and sons presumably decided to walk away from this (relatively marginal) holding.

Today Joseph lives in the former farm headquarters on a small hill west of the Ngare Nanyuki River, surrounded by the compounds and fields of his wives and children. The Momella squatters, excepting Molel and Mzee Lesura, moved to other areas of the Ngare Nanyuki Ward in the 1960s, buying partial shares of the settler estates that were subdivided among Meru families after independence. Molel stayed in his compound on the east side of the river and, with his four wives, raised twenty-six children. Collectively, the offspring and grandchildren of Joseph and Molel make up a substantial portion of Nasula's population and have a significant role in local politics. For example, one of Molel's sons was Olkangwado Village chairman during the period of this study, and others were in prominent positions in the local church and the *kitongoji* committee.

The history of Nasula lands as former settler freeholds and as former squatter settlements, in combination with local ecological conditions, has created a socially and spatially complex pattern of land tenure. To begin, the lands of Nasula gradually slope downward in a northerly direction away from the mountain. Much of the northern section of the former Monas estate is composed of seasonal swamp, bisected by the Ngare Nanyuki River running south to north. Much of this seasonal swamp is treated as a dry season grazing commons and is used by several villages, including Olkangwado, Lendoiya, and Leguruki. A small portion of it, mostly on the west side of the river, is treated as a more exclusive dry-season reserve for Nasula Kitongoji residents only. The area is staked off with tree branches and actively patrolled to keep all livestock off from May to September. The Trappe estate portion of Nasula was for the most part controlled by Molel and another early settler, Mzee Lesura. In this area, individual households have exclusive rights to the land immediately surrounding their compound. The land between compounds and the open pastures surrounding the schoolhouse and beyond, up to the park boundary, function as commons. In the portion of the *kitongoji* surrounding Monas's former farm headquarters, a similar pattern of land use and tenure prevails. In this area, it is the steeper slopes of the mountain that constitute the grazing and bush commons. Throughout the *kitongoji*, households control exclusive rights to cultivation plots scattered at distances varying from one to six kilometers from the family compounds.

The two sections of Nasula, the Monas estate and the corner of the Trappe estate, operate under slightly different moral economies. The moral economy of land on the former Trappe holding functions close to the ideal for *kihamba* lands. Molel, Lesura, and men like them in nearby Meru expansion areas such as Lendoiya and Leguruki are viewed as pioneer settlers who have staked extensive claims on uninhabited land.[53] The claims were large enough to support multiple wives and to provide sons with their rightful inheritance. On the Monas estate, however, the moral economy of land operates differently from *kihamba* lands, in large part because of the way it was transferred to Joseph. To be sure, Joseph is identified as the family patriarch and his sons inherit their portion of his *vihamba* lands. Joseph's role, however, is not restricted to that of pioneer patriarch. When Joseph received his "pension" in land, his social identity was altered. He suddenly had control over an extensive property, one much larger than a pioneer settler could claim and control if he were clearing uninhabited bush. He was not seen as the individual "owner" of this property, but rather, was and is viewed locally as a powerful patron who can grant access to land and pastures to those in need. The extensive grazing commons in the seasonal swamp is seen as a collective resource, not merely for Nasula but for several villages. Within the local moral economy, it is inconceivable that Joseph would fence off access to this area or charge rent in the way that Monas did. It was also inconceivable that he lay exclusive claim to all the arable land of the old estate. Joseph claimed some of the best lands for himself and his wives and then granted portions of the remaining lands to those who solicited his patronage. In the years immediately following independence, in-migrants wishing to settle on estate lands obtained rights of occupancy by bringing gifts of *pombe* (beer) and livestock to Joseph, who would grant them a site.

It has been many years since access to land has been granted in this manner. In the survey I conducted, the only immigrants since the late 1960s whom I identified were women who had married into the village. There has been very little in-migration into the area from other parts of Meru in recent years, as most lands have long been claimed as individual plots or *kitongoji* commons for grazing and fuelwood reserves. As population increases, the areas of common pasture and bush are being converted to cultivated plots, repeating the historical pattern observed in central Meru. Only a quarter of those surveyed responded that they had insufficient land for farming, but 100 percent responded that it will be difficult to find land for the next generation. Observed one resident, "There is a problem for the young men to get land since we are surrounded by the national park and other areas are already occupied."[54]

Many people are land poor, but as yet there is no distinct landless class in Nasula. A strong subsistence ethic (Scott 1976; Watts 1983) still operates within the *kitongoji*, and everyone has access to a cultivation plot, no matter how marginal. As in similar moral economies (see, e.g., Feierman 1990) community members who have no other access to land are allowed cultivation plots on the less productive lands. In Nasula, where most of the land is relatively marginal for cultivation (see below), it is the pasture and bush commons close to the national park boundary that are least productive—not because of inherent ecological characteristics, but because of the susceptibility to predation by wild animals coming from the park. In this area of the *kitongoji* are the homesteads of divorced and widowed women and, increasingly, the homesteads of newly established households. On both sides of the Ngare Nanyuki River, the bush and pasture commons are being progressively divided among individual households and converted to cultivation plots (see figure 4). Though land scarcity is part of the motivation for this phenomenon, as we will see in the final chapter, there is also a collective desire to increase tenure security in this area as a way to counter the expansionist designs of the national park administration.

The land use patterns in Nasula reflect the complexity of tenure. Unlike in Akheri, cattle are grazed on common pastures, not stall-fed. Historically, the Ngare Nanyuki Ward, to which Olkangwado Kijiji and Nasula Kitongoji belong, was an important Meru grazing area, and it continues to be used as such. It is not uncommon to see herds of over fifty cattle on Nasula's common pastures. A strong tradition of cattle keeping persists in the habits of elders like Molel (whose own herd was said to number sixty head). Large herds are losing their symbolic importance as wealth and security, however, and Molel is viewed as somewhat of a throwback. People increasingly stress the importance of keeping fewer, more productive livestock. Nevertheless, there are few pure or mixed Jersey or Frisian cattle, largely because of the prevalence of livestock diseases and the scarcity and high cost of veterinary medicines. In fact, cattle diseases are viewed locally as a major constraint on livestock keeping. Ninety-three percent of the survey respondents reported that they were now keeping fewer cattle than before, not because they own better breeds, but because of disease losses and the growing scarcity of pasture.

The majority of households own one or two cows and a mix of smaller livestock. Except for in the small dry-season reserve and cultivated plots, livestock are allowed to graze virtually anywhere in the *kitongoji*. Commonly five or six families graze their cattle together and rotate herding duties on a weekly basis. Most of the large herds of livestock seen on the common pastures are not owned by a single person,

Figure 4. A new homestead cleared out of the Nasula Kitongoji bush commons. (R. P. Neumann.)

but are herded together as a more efficient use of labor. Much of the routine daily herding around the *kitongoji* is left to children ranging roughly from six to sixteen years of age. One of their most critical duties is to prevent livestock from wandering into cultivated plots or crossing the national park boundary. Donkeys, sheep, and goats often wander untended among the cluster of houses during the daytime. At night, livestock are kept within the compounds, either in kraals or mangers.

The spatial pattern of cultivation in Nasula also differs from that in Akheri. There are relatively few compound gardens in the *kitongoji,* and cultivation plots tend to be fragmented and scattered at various distances from the houses. The rocky and shallow soils of Nasula, described in greater detail in chapter 5, are much less productive than those of central Meru. Consequently, much cultivation is conducted in the small depressions of the *kitongoji's* hummocky terrain on the east side of the river, where pockets of relatively fertile soil are found (see figure 5). Within the river's floodplain, there are also relatively deep soils that are cultivated in contiguous plots averaging about .8 hectare

Figure 5. Goats graze on crop stubble in a tiny pocket of bottomland in the hummocky terrain of Nasula. (R. P. Neumann.)

per household. In recent years, villagers have also begun to divide and cultivate .4-hectare plots on the mountain slopes on the west side of the river, in what had been common bush and grazing lands. Many households also have plots outside of Nasula, several as far away as Kisimiri, a six-hour walk. All of these areas of cultivation are rain fed. Access to irrigated land is rare and limited to an area below the old Monas farm headquarters fed by a small trench coming from the upper slopes of the mountain. Fifty-five percent of the survey respondents said they had access to small parcels of irrigated land, most ranging in size from .05 to .1 hectare.

Production is mostly oriented toward household subsistence. There is no mechanization at all and ox plows remain the principal means of cultivation, though many residents are too impoverished to own oxen or hire them for plowing. Conditions are not conducive to coffee cultivation in most of Nasula, and virtually no one plants it. Most farming is dedicated to the cultivation of food crops such as maize, millet, beans, potatoes, sweet potatoes, and yams. Those few residents who have access to irrigated land can grow bananas and an extra maize crop, as well as local cash crops like tomatoes, cabbage, and onions. In recent years, a rather volatile market for tomatoes emerged when Chagga merchants began to travel through the area to buy produce. Irrigated land appears to be increasingly used for tomato production, thus helping to push subsistence crops onto common pasture and bush lands.

The Ngare Nanyuki Ward is, in sum, the most underdeveloped area in Meru. The inaccessibility of the area makes agricultural and livestock inputs expensive or unavailable altogether. The weekly village market about one and a half kilometers to the north in Olkangwado and the Chagga traveling buyers are the principal outlets for surplus crops. Transportation is far too expensive to allow crops to be marketed in Arusha directly by producers. Aside from the two motorcycles owned by the ward chairman and a Meru Baptist minister, and a pickup belonging to a retired government official, there were no vehicles in a ward of over nine thousand people. There are no electricity or phone lines within ten kilometers. However, a water line supplying Kilimanjaro International Airport many kilometers away passes through Nasula and provides residents with high-quality drinking water.

LAND AND SETTLEMENT IN NGONGONGARE

Ngongongare Kitongoji lies on the southern boundary of Arusha National Park and is part of Ngurdoto Village. Tanzania's 1988 national census (United Republic of Tanzania, 1989) recorded 5,008 people living in Ngurdoto, an increase of 47.5 percent from the 1978 total of 2,627. Like Nasula, the *kitongoji* was settled by people migrating from the more congested villages of Meru, such as Ndoombo, Nkoaranga, Songoro, and especially Kilinga. Unlike Nasula and Akheri, a substantial number of people have migrated from more distant areas of land scarcity, such as Mount Kilimanjaro and the Pare Mountains in northeastern Tanzania. As in Nasula, the land that the *kitongoji* now occupies was originally alienated by the Germans, and the parcel was operated as a coffee farm and cattle ranch by European settlers well into the independence period. And like Nasula, Ngongongare Kitongoji was settled by Meru livestock keepers who worked on European estates to gain access to grazing pastures. Mzee Perimu recalls coming from Mlala in 1960 with about ten other people: "We came and asked for a plot from a European, Figenschou, to graze our cattle. He gave us a plot just to graze our livestock, and then we would go back to

Mlala. Clearing the bush on his land was the payment for grazing our own livestock. When TANU became harsh [*kali*] about his holding, he decided to leave, and so the people who were grazing there got a chance to settle the area and cultivate."[55] The estate was confiscated by the government in 1969 to settle the owner's debt, and a portion of it remained in possession of the grazing tenants and others who followed. The remainder of the parcel was eventually gazetted as part of Arusha National Park.

Ngongongare, like Ngare Nanyuki, was used by Meru herders for grazing before European conquest, sometimes with Maasai. An ancient cattle track led through the heart of the alienated parcel that linked central Meru with lowland plains to the south and highland plains to the north, and so with surrounding Maasai pastoralists, in a trading network that extended to Kenya and the coast. Mzee Pallanjo remembers that the control of the pastures flanking the cattle track was a point of struggle between Meru and Maasai herders: "Long ago this area was controlled by the Maasai and a few Meru . . . [O]ur ancestors were living in this area keeping cattle, but after tribal wars they decided to shift their homes to the other side, to Meru, and then just come down to graze here . . . After a time the Maasai found that the area was not good for their needs and they migrated and left the area for the Meru."[56] The reference to the Maasai's withdrawal is probably related to the rinderpest epidemic and droughts of the late nineteenth century, which devastated livestock populations in this region (Spear 1996). Thus when German pacification forces arrived, they likely found only a scattering of Meru herders in the area, and the government proceeded to alienate the land for European settlement.

Ngongongare is the last area on Mount Meru to be opened up for settlement by people pushed from the congested lands of central Meru. Apparently, the estate owner was delinquent in mortgage payments and taxes for the property in the late 1960s, and the Tanzanian government nationalized the land. This afforded the nonresident Meru herders an opportunity to move onto the land in 1969 and 1970. Mzee Perimu, one of the original grazing tenants, remembers that local government officials tried to halt the settlement of the property: "[Figenschou] decided to leave everything behind, and with this the people who had been grazing there saw an opportunity to invade the area and cultivate . . . [T]he Meru Council at that time tried to clear the people off, threatening if they [didn't] move they were going to burn their houses. But they didn't dare and we have remained up to this time."[57]

The TANU government did, however, manage to stop the new settlers from occupying the entire estate by extending the boundaries of Arusha National Park. The park staff defended the new park boundary against new settlers until it was officially gazetted in 1973. The new park extension thus came to border the new cultivation areas. Within the park boundaries, the coffee trees and other traces of European occupation of Ngongongare Farm were removed.

The history of how the Ngongongare came to be occupied and the spatial relationship to the national park boundary together produce an interesting pattern of land tenure. The alienated block of land that constituted Ngongongare Farm was surrounded on three sides by the Meru Forest Reserve, originally gazetted by the German colonial government. The *kitongoji* lands consist mostly of undulating plain gradually rising toward the north and more steeply to the forested slopes to the west. The people who made up the first wave of immigrants operated as pioneer settlers moving into an area of uninhabited bush. They cleared for themselves as much land for cultivation and livestock as they could control. As a result, all of the land became divided up into individual household claims. The only commons left was the plot of land set aside for the Ngongongare primary school. Nothing equivalent to the village bush and pasture lands of Nasula exists for the expansion of new homesteads.

The moral economy of *kihamba* still operates in the *kitongoji*, with sons inheriting from their fathers when they reach the age of marriage. Among the oldest of the pioneer settlers, land has already begun to pass to the next generation. Eighteen percent of the survey respondents reported that they had obtained their plots through inheritance. As yet the kinds of intergenerational tensions over land found in Akheri have not emerged in Nasula, though virtually everyone recognizes that the coming generation will face some difficulty. None of the households in Ngongongare are landless, but because there is no land left for expansion, some people find themselves with inadequate access. Nevertheless, the subsistence ethic is still maintained. Plots for cultivation are commonly borrowed and lent on a seasonal basis without any formal payment of rent. Thirteen percent of the respondents reported that they commonly borrowed land from friends and relatives.

As the population figures imply, people have continued to migrate into the *kitongoji* from elsewhere. Forty-one percent of those surveyed had settled in Ngongongare since 1980, the year of the previous national census. Over half of these newcomers had come from outside of Meru, including Mount Kilimanjaro and the Pare Mountains to the east. The recent arrivals from outside of Meru have obtained access to land by purchasing plots from the original settlers. Thus, despite the continued prevalence of the *kihamba* ideal, a limited land market exists in Ngongongare. The market is limited geographically, as the plots that have been sold are only .2 hectares in size and are strategically located along the boundary of the park. In a later chapter, this pattern of tenure and subdivision will be discussed in the context of village-park conflicts. Since the land sales are small and next to the park boundary, they as yet pose no challenge to the moral economy of *kihamba* land.

The *kihamba* plots in Ngongongare are an attempt to reproduce the traditional Meru ideal, wherein a household controls enough land to both raise cattle and cultivate (see figure 6). One respondent's explanation that "I decided to shift to this area in order to get enough land for farming as well as raising cattle" is a typical expression of newcomers' hopes.[58] The spatial arrangement of household compounds, cultivated plots, and grazing pasture is thus quite different from that in Nasula. Typically people live in homesteads whose boundaries contain house sites, cultivation plots, and limited grazing areas. The greater share of household production takes place within these holdings. Nearly half of those surveyed (45 percent) said they farmed and raised livestock exclusively on their land surrounding their compound. Most of the other respondents had only one additional plot, typically located in another *kitongoji,* at distances ranging from two to ten kilometers away. The houses, then, are dispersed and widely spaced, with cultivated fields (usually bounded with living thorn fences) and pastures lying between. Most household compounds have intercropped fields, livestock stalls, and kraals close in, with pastures at further distances. Virtually all of the households plant various combinations of maize, beans, bananas, and sweet potatoes.

The *kitongoji's* proximity to a main transportation route facilitates greater household involvement in external commodity and labor markets. Increasingly, coffee is intercropped with bananas and other staples much as it is in Akheri, though without irrigation. Coffee has been cultivated in Ngongongare Kitongoji as early as 1976, and by 1990, 45 percent of the survey respondents reported that they were growing coffee. Household production in Ngongongare is thus more market-oriented than in Nasula. There is a coffee-marketing cooperative in Ngurdoto Kijiji center where growers can take their beans, and the Moshi-Arusha tarmac road is only a ten-minute drive on a maintained

Figure 6. Plowing the maize plot at a homestead in Ngongongare. A cattle stall and banana garden are seen in the background. (R. P. Neumann.)

gravel road. Using donkey carts, handcarts, pickup trucks, and bicycles to get produce down to the bus lines, producers are able to market their own crops in Arusha. The production of surplus food crops, however, is not a significant source of household income. The main economic advantage of the village's location is access to wage labor. Twenty-three percent of survey respondents reported that their largest source of income for the household was salaries from employment outside of the village.

Large herds such as are seen in Nasula are absent in Ngongongare. Since no common pasture outside of the primary-school grounds exists, the size of villagers' livestock holdings are constrained by the size of their individual plots. Goats and sheep are herded on household pastures and occasionally on the school grounds. Almost all households keep at least one cow, often of a mixed or improved breed such as Jersey. The inoculations and other treatments necessary to raise improved breeds can be obtained much more easily in Ngongongare because of its proximity to the tarmac road. The location also allows easier marketing of surplus milk, and 27 percent of the respondents reported

Figure 7. A Ngongongare farmer and his two sons bring maize stalks to the cattle stall near their house. (R. P. Neumann.)

income from milk sales. Increasingly, cattle are at least partially stall-fed, and supplementary fodder such as maize stalks and banana leaves is brought in from cultivated plots (see figure 7). Still, most residents find it extremely difficult to obtain enough feed for their livestock, and loss from disease is a major problem. In sum, though Ngongongare is a bit more prosperous than Nasula, the village has no electricity or running water. The area is less drought prone than Nasula, and the soils deeper and more productive, but large areas are not well drained and many of the low-lying areas are marshy. All agriculture is rain fed and mechanization is rare, with most plowing done by animal traction.

The political and social context within which Arusha National Park eventually emerged is marked by nearly a century of conflict over land between the state and Meru peasants. The loss of access to a large portion of their resource base meant the breakdown of their production strategies and land tenure system and a crisis of land scarcity that intensifies as populations increase. Questions of land and resource access have been central in the politics of Mount Meru since the beginning of the colonial period.

The dislocations of the past continue to be important in shaping political consciousness and action on Mount Meru today. The geography of settlement and land use on Mount Meru serves as a daily reminder of the colonial alienations for settler farms, the establishment of the forest and game reserves, and the eviction from Ngare Nanyuki.

The political tensions created by the scarcity of land on Mount Meru were accompanied by tensions in the local moral economy. The strains in the moral economy are expressed in many forms, from confrontations between fathers and sons to debates over sharing and reciprocity. Though a subsistence ethic remains the ideal in many villages, tensions surrounding its application in practice remain. For example, in times of seasonal dearth, it is a common survival tactic of poorer community members to go "visiting" during normal meal hours. The customary response of the host is to invite unannounced guests to join the family in their meal. The "haves" of the villages complain that their food stores are disappearing from such visits and debate their responsibilities toward the "have-nots." Increasingly, younger, more educated people with access to income outside of the villages deny any responsibility to feed uninvited visitors. Additionally, in villages such as Akheri where further subdivision of *kihamba* land is becoming untenable, the guarantee of land access has broken down, and landlessness is growing common. Thus the terms of local moral economy are shifting according to the level of involvement in outside labor markets and the possibilities for expansion into uncultivated land.

The historical land crisis continues to influence the contemporary relationship between Meru peasants and the state. In particular, Meru response to the Ngare Nanyuki eviction demonstrated that a culturally and economically differentiated society could unite against the colonial state in a common defense of land rights. Tanzania has been independent of British rule for over thirty years, but most of the colonial alienations remain intact and the state persists in making land use decisions contrary to the desires and interests of the peasantry. Disenfranchised after independence, peasants have resisted "the government of the bureaucrats just as their predecessors had resisted the government of the colonialists" (Feierman 1990, 26). On Mount Meru, the historical pattern of resistance to the loss of land and resource access continues to guide local responses to contemporary nature-protection laws and policies.

3

Conservation versus Custom

State Seizure of Natural Resource Control

---

*"[P]reservation" of wild life as "game", was directly and
repeatedly challenged by men living and finding their living
in their own places, their own country, but now, by the
arbitrariness of law, made over into criminals, into rogues,
into marginal men.*

> Raymond Williams, *The Country
> and the City*

*The demarcation of the Forest Reserves on Kilimanjaro and
Meru and the prohibition of cultivation and grazing within
the boundaries was probably as unpopular a thing as the
government ever did in those parts.*

> Tanganyika Territory, *Report of the
> Arusha-Moshi Lands Commission*

Securing control over access to, and the benefits derived
from, natural resources was a critical process in the early formation of
the colonial state in Tanzania. Natural resource laws were essential not
only for generating revenue for the state and fueling accumulation for
private interests, they were symbolically important for the assertion of
the dominance of the German kaiser and later, the British Crown, over
all aspects of the territory's economy and wealth. The resulting cen-
tralization of control was produced at the expense of an existing system of communal property relations and customary rights to land
and resources within African societies. The purposes of this chapter are
to identify the ways in which the new colonial dispensation disrupted

97

customary practices and rights and to explore the ways in which Africans resisted these changes.

Much of the chapter is based on archival sources found in Tanzania and England, with additional information from various Tanzanian government documents.[1] Relying on these sources for insight into the desires and motivations of Africans under colonial rule poses analytical challenges. Since the written history of conservation is the product of an elite social group, the voices of African peasants and pastoralists are heard here as barely audible whispers, and even those are usually relayed secondhand. These types of documents, written by state officials, have limited utility for uncovering "the silent and anonymous forms of class struggle that typify the peasantry" (Scott 1985, 36). Nonetheless, we can gain a sense of what was at stake for rural African societies by tracing some of the debates conducted within the colonial government concerning the conflicts between conservation policies and customary rights to land and resources. Occasionally, the actions of those whose land uses were threatened by conservation policies are reported in the colonial records, and these incidents hint at the existence of a rural moral economy, its constitution, and its defense. The chapter, then, examines the historical process of transference of natural resource control from the local customary institutions to the state, beginning with general patterns in the territory and moving on to the specific situation on Mount Meru.

Natural Resource Control and the Colonial State

State forestry under German rule began slowly in 1892, gaining momentum in 1903 with the appointment of the first full-time professional forester and the enactment of the Forest Conservation Ordinance a year later (Schabel 1990). The ordinance created a system of forest reserves and established prohibitions against their use. According to Schabel, German motivations for establishing reserves were more environmental than fiscal. Nonetheless, the Germans were interested in making the territory profitable and did seek to develop timber production for both domestic and external markets. Ultimately, however, timber would not contribute to colonial coffers, operating expenses remaining about double the revenue for the duration of German rule

(1990). In fact, most of the German forestry officials' energy and finances were directed toward the exploration, demarcation, and survey of forest reserves. A visiting forestry expert commented in 1935 that "[b]etween 1896 and 1914 this work was pushed on energetically"[2]— nearly an understatement considering that the Germans had proclaimed 231 reserves from 1906 to 1914.[3] The effect of German forest laws on existing African access and use was direct and immediate. "Under Teutonic discipline"[4] (which included corporal punishment and confinement in chains), all African settlement, cultivation, burning, and grazing was outlawed in designated forest reserves.[5]

The Germans made similar efforts to control wildlife, legislating a complex set of regulations on hunting, as well as creating eighteen game reserves where all hunting was prohibited.[6] A hunting license was required to hunt most animals in the colony. Africans also had to have a game license to hunt any of the controlled species, which included common meat sources such as antelopes, buffalo, and hippo. The only animals anyone could hunt without a license were predators such as river pigs, warthogs, porcupines, ground pigs, and monkeys.

World War I took its toll on the German forest bureaucracy's meticulous record-keeping efforts in German East Africa, and British natural resource professionals thus found that few of their documents remained.[7] Enough was preserved, however, for the British to use as a base on which to build their program of resource management and conservation. Regarding the draft of the first Regulations on the Conservation of Forests, the interim director advised, "I cannot do better than to refer you to the laws in existence under the German regime." The British government, he continued, "would be well advised to base its forestry laws on those of the Germans."[8] Essentially it did, immediately proclaiming all reserved forests anew as a preliminary measure.[9] In December 1920, D. K. S. Grant (previously of the Kenya Forest Service) was appointed the first conservator of forests. His primary charge upon taking office in January 1921 was the creation of a separate Forest Department based at the old German forest headquarters at Lushoto. The legal framework for administering the territory's forests was established by the 1921 Forest Ordinance, which incorporated all the previously designated German forest reserves. Gazetted forests in 1921 totaled 8,770 square kilometers, slightly less than 1 percent of the territory.

Once the new Forest Department assessed the forests in its charge, some were decommissioned, others added. By 1925, 212 reserves covered 9,601 square kilometers, most of which were closed montane tropical forests in the highlands. The 1921 ordinance initiated a series of prohibitions for these reserves, including cutting or removing trees or forest produce, firing, squatting, grazing, and cultivating.[10] As restrictive as these rules appear, the policy was such that it did not prevent "the exercise of any right or privilege recognized by the Governor," whose officers could issue licenses for most of the prohibitions. One significant (and contentious) concession was the free use by Africans "of any forest produce taken by them for their own use only."

As with forest resources, the British relied heavily on the work of the Germans in establishing state control over wildlife. The Game Preservation Ordinance of 1921 repealed the German Game Ordinance of 1908–1911 and more or less regazetted German game reserves (see table 2). The intention was eventually to discard and replace those that did not fit future plans for the territory and to gazette new reserves as necessary.[11] There were three types of reserves, and of these the "complete game reserve" had the strictest management. No hunting was allowed and the governor had the power of "prohibiting, restricting, or regulating" entry, settlement, cultivation, and the cutting of vegetation. Even at this early stage of British administration a substantial amount of land was allocated for wildlife conservation (see map 5). Successive game wardens were particularly keen to discourage all African settlement in the reserves.[12]

Besides the establishment of reserves, the 1921 ordinance contained clauses regulating African hunting which were important for customary rights. The law established that "No person shall hunt any game unless he holds the appropriate Game License" and that "No license shall be issued to a native without the consent of the Governor." The regulations (Section 3[3]) established under the ordinance further outlawed certain traditional hunting practices, including the use of "nets, gins, traps, snares, pit-falls, poison, or poisoned weapons." On the face of it, the law seems to be quite restrictive in regard to African hunting, yet relative to those of other colonies, it was in fact liberal. The governor could make regulations "allowing and regulating the hunting of game for the purpose of food supply in times of famine or by natives who are habitually dependent for their subsistence on the flesh of wild animals" (Section 3 [e]). In fact the policy of the government was such "that the native should be regarded as having a moral right to kill a piece of game for food."[13]

Conservationists in England—most notably the Society for the Preservation of the Fauna of the Empire (SPFE)—found this "liberal"

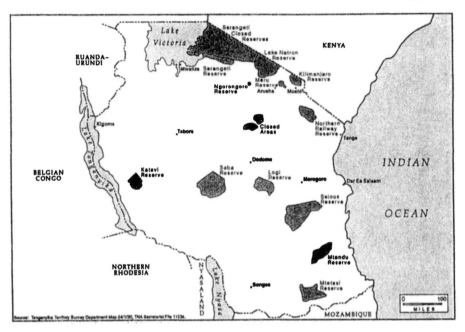

Map 5. Complete and closed game reserves in colonial Tanganyika, 1930. (Tanganyika Territory Survey Department Map 24/1/30, TNA 11234.)

policy unpalatable and constantly chastised the Tanganyika government for allowing the "slaughter" of game by Africans. As an indication of the SPFE's influence, the Game Ordinance of 1940, which replaced the 1921 ordinance, was authored "to give effect to the provisions of the International Convention signed in London on the eighth day of November, 1933, in so far as those provisions relate to the preservation of fauna in its natural state."[14] The London convention resulted almost exclusively from the SPFE's efforts.

The 1940 bill, however, did not significantly alter Tanganyika's game policy, with the exception that a new category of protected area, the national park, was created.[15] Contrary to what was sought by European conservationists, the wording on African hunting seemed, if anything, to more strongly recognize traditional rights: "Nothing in the foregoing provisions of this Part shall make it an offense for a native to hunt, without a license, any animal not protected under the provisions of section 24, for the purpose of supplying himself and his dependents with food, provided that he does not use arms of precision."[16] As for game reserves and national parks, certain customary rights were also confirmed in the 1940 Game Ordinance. Any "person whose place of birth or ordinary residence is within the reserve" or "who has any rights over immovable property within the reserve" could enter or reside within. Immovable property, it was ruled, included pasturage for livestock.

The stage was thus set for a protracted battle between conservationists and human rights advocates over the configuration of wildlife and forest management in the colony. On one side, conservation advocates wanted the state to exercise its claim as the sole legal authority controlling natural resources to curtail any customary use by Africans. On the other side were pro-African elements who argued that the League of Nations mandate obligated Britain to respect African rights to the greatest degree. The result was a policy of compromise: not a compromise between the state and the colonized African population, but between factions within European society, and one that was ultimately unworkable given the ideological chasm dividing the two sides.

Conservation versus Customary Rights

The legal ambivalence of colonial policy and legislation allowed both African rights and conservation advocates alike to make convincing arguments about the validity or irrelevance of African customary claims to land and resources. Very early in the British occupation, Sir Donald Cameron, governor of Tanganyika from 1925 to 1931, foresaw the opposition of pro-African officials to the game laws and suggested that "the interests of the people must be paramount and that the conventional attitude as regards game preservation requires revisions."[17] In particular, agricultural and veterinary officers thought that nature preservation efforts lacked discrimination and too often had the effect of impoverishing rural areas. A resolution passed at a 1926 agricultural conference stated: "[I]n Tanganyika . . . indiscriminate game preservation . . . had the effect of so segregating natives that their land was becoming exhausted, and a condition was arising leading to their demoralisation and preventing their natural increase."[18] The observation is an amazingly prophetic statement about the conditions surrounding protected areas on the eve of the twenty-first century.

This resolution and similar criticisms of game and forest policies were hotly contested. The chief native commissioner responded that he knew of nowhere that game preservation had crowded Africans "in areas which are too small to carry them." If anything, preservation promoted development "by dissipating the natives who prefer to wander abroad and to spend on hunting game time and energy which would be more properly devoted to agriculture."[19] For their part, the wildlife and forestry professionals simply refused to recognize that being an indigenous African in a colonized territory afforded any special treatment in regards to land and resource access. The conservator of forests reasoned that since intertribal warfare kept territorial boundaries in flux, there was no basis to acknowledge that the African's "claim is more valid than that of the non-native."[20] Commenting on the proposed forest rules of 1928, one official decreed that "[t]he natives have no more inherent property in the forest than in the land, and, besides, must always be protected against themselves."[21] The colonial government's rulings at times seemed to support such positions, as in the attorney general's 1926 decision that all lands, whether alienated or tribal, were public lands and the government therefore owned *all* rights to cut and sell timber from African lands.[22] Yet even such clear-cut legal decisions were disputed at the highest levels. "I entirely dissent from the view," wrote Governor Cameron in 1927, "that the Forestry Department, is in any circumstances, entitled to credit in respect to royalty on timber on which they have not expended time or money and in most cases have never seen."[23] Nevertheless, foresters singled out policies favoring customary rights, such as free issue (forest products collected on government forest lands without payment of a fee) for African household use, as a major impediment to fulfilling the goals of scientific forestry.[24]

The debate over customary rights versus scientific forestry swung decidedly in favor of the latter after an advisory visit by Professor R. S. Troup, director of the Imperial Forestry Institute in 1935. The recommendations of the Troup Report guided forest policy in Tanganyika for nearly twenty years. In his report, Troup presented a lengthy polemic on the evils of free issue. He began by citing Forest Department estimates that the annual royalty value lost to free issue from 1923 to 1934 averaged £12,786. These figures were intended to show the potential surplus to the department if free issue were eliminated. For Troup, free issue was a hidden subsidy for Africans. He felt the policy of allowing Africans free use contradicted the Forest Ordinance, which implied that reserved forests are free of rights. He concluded, finally, that "the timber trade of the country is handicapped by free issue . . . So long as natives are allowed free produce from forest reserves, the development [of a trade in small dimension timber] will be difficult." He recommended that the government "declare the non-existence of any rights to free produce from existing forest reserves" and levy royalties on any and all forest products.[25]

Such opinions and debates indicate how dangerously ambiguous the various land and natural resource laws were. An untenable situation existed in which African rights seemed to be simultaneously eliminated and protected. A district administrator, for instance, argued that the Game Ordinance opposed the Land Ordinance and Territorial Mandate because it "must necessarily interfere with the holding, use, occupation and enjoyment of lands by natives and must necessarily disregard the rights and interests of natives."[26] These contradictions are representative of a pattern in the colonial state's approach to land and resources laws. In asserting its political dominance, sweeping claims to the ownership of all land and resources would be made and then the "privilege" to continue *some* customary uses would be granted. The character of these privileges was shaped by the colonial political economy as well as European ideologies of nature, hunting, and scientific resource management. Both of these factors will be explored below.

Some colonial officials, such as the "pro-African" administrator Sir Philip Mitchell (later governor of Kenya) and his onetime protégé in Tanganyika's Northern Province, A. E. Kitching, seemed sincerely concerned with protecting customary rights against the onslaught of European conservation efforts. Kitching's articulate and impassioned counterattacks on forestry and wildlife officials in defense of African rights appear in the records repeatedly, sometimes as a lone voice. Many of Kitching's critiques were carefully reasoned legal arguments based on existing laws. Other administrators chose to argue from a moral stance. "Are we justified," asked an official, "in treating a man who, following the customs of his ancestors, keeps a bee-hive in the forest without a license or searches for wild honey as a criminal who may be sent to gaol for six months?"[27] Moral appeals and outspoken advocates like Kitching were rare, however, and the harsh political-economic realities of administering an insignificant colony such as Tanganyika go further in explaining the government's reluctance to extinguish African rights than some model of imperial benevolence.

To begin with, Tanganyika was never regarded by Britain as an important territory. The amount of money and effort Britain was willing to invest in order to control its territorial mandate was minimal, and the Game Department was far down on the list of priorities within the administration. In fact it was nearly eliminated in the budget cuts of 1931, a cheering notion for many provincial commissioners.[28] In short, though it claimed sole ownership of the territory's land, wildlife, and timber, the state's ability to completely control the rural areas was limited. The Chief Secretary of Tanganyika, in fact, admitted in the public press that the government could not control African hunting and was taken to task by the SPFE in London.[29] African game scouts, often the sole representatives of the state for many kilometers, were heavily dependent for their welfare on the communities whose members they were supposed to be arresting, and "bribery" was widespread.[30] For some administrators, the proposal to impose fees for Africans' use of forest products was irrational since it would be impossible to enforce without incurring great expense.[31] Benevolence was cheaper.

In addition, administrators were well aware of the importance of free access to resources to rural livelihood and that if they pushed so far as to threaten survival, rebellion might ensue. In the aftermath of the Maji Maji Rebellion of 1905, for instance, German military officers speculated that African opposition to the game laws had helped trigger the uprising (Koponen 1995). While foresters urged the implementation of a fee schedule for minor forest products, administrators warned that the "proposal would arouse intense opposition" among Africans.[32] Of all Troup's recommendations, the elimination of free use for Africans drew the most cautionary responses. Governor Harold MacMichael warned that it was not "wise to make any sudden and drastic change" in policies concerning African forest access. For administrators concerned with order and stability and a modicum of legitimacy, the proposals from natural resource professionals to curtail customary rights were political red flags.

Finally, free access to subsistence resources helped to keep African wages depressed, since essentials such as fuelwood and building poles could be obtained by the labor of wives and children rather than by cash purchases. Furthermore, in rural economies, game meat and honey collected in the forests were critical nutritional sources. Hence, the interpretations or explicit wording of wildlife and forest laws directed that Africans could take what they needed for their own consumption. At the same time, the laws prohibited them from using their access "privileges" to participate in the market economy by selling meat or timber. In this way natural resource management laws helped push Africans toward wage labor to meet their cash needs and prevented them from competing with Europeans in the marketing of wildlife and forest products. In short, allowing Africans free access left them with one foot in the subsistence economy and helped fuel the accumulation of capital among Europeans.

Colonial resource policies were indeterminate and ambiguous about African rights because political-economic conditions were such that they could not be otherwise. In essence, the law placed all regulatory control in the hands of the governor, while granting some African rights to hunt and collect products of the forest. Consequently, the government could avoid the political trouble of attempting to curtail all African rights without having to explicitly relinquish state authority. Customary African hunting and land use practices, if not redefined as "crimes," were considered "privileges," theoretically revocable at the discretion of the governor. However obtrusive this situation was for Africans, it was far from satisfactory for the conservationists in London.

THE UNHAPPY HUNTING GROUNDS

Reflecting on the controversies and conflicting interests at play in the debates surrounding the 1940 Game Ordinance, Tanganyika's solicitor general remarked that "[t]here can be few such subjects which are capable of rousing such violent passions in the breasts of otherwise quite reasonable people than that of game preservation."[33] While the colonial administration was willing to concede limited local access, various other interests within European society were anxious to see it extinguished. African rights to hunt wildlife were perhaps the most visible and symbolically important arena of struggle in this respect. The culture of "the Hunt" (MacKenzie 1987) was deeply imbedded in the consciousness of European settlers and resource professionals, especially those with military backgrounds. Hunting wild animals in Africa was an important symbol of European dominance on the continent and instrumental in distinguishing social class within settler society (MacKenzie 1987; 1988). Success in hunting translated to success in military campaigns, at both the individual and group level.

Hunting in Africa was ultimately transformed into a ritualized act that became one source of upper-class identification, a measure by which a social group could differentiate itself from other classes and, in particular, Africans (MacKenzie 1988; Neumann 1996). A complex set of mores and rules concerning the killing of wild animals were established which were embodied in the terms "sportsmanship" or "sport hunting." A person's class status determined, and could be determined by, how they brought an elephant to its untimely demise. The idea of hunting with the sole motivation of obtaining meat was thus anathema within the ideology of the Hunt. For the European engaged in the hunting ritual, food acquisition was largely irrelevant. For African societies engaged in their own ritual hunting practices, efficiency in obtaining animal protein was highly valued.

The discourse of colonial conservation thus soon became centered on the "unsportsmanlike" quality of African hunting practices. Conservationists singled out Tanganyika as the worst offender,[34] mostly because its policies were relatively sympathetic to African hunting rights. For example, at a game conference in Mombassa in 1926, Tanganyika's game warden explained that the policy is "as regards natives, to take no notice of minor or isolated offenses."[35] A subsequent regional conference on the protection of wildlife held in Nairobi in 1947 became a forum for attacking Tanganyika's policies. Conference participants deplored the tolerance of African hunting methods in moralistic terms— what was "game" for Europeans was a "victim" for Africans. Regarding the use of poisons, a critic declared: "[T]he victim [animal] dies slowly in sweating agony" under the "foulest and most unsporting method of killing."[36] Another participant declared that "there is wholesale extermination [of game] by native hunters" in Tanganyika.[37]

People outside the colonial bureaucracy also pressured the government with denunciations of their game policy. Editorials by such public luminaries as Sir Julian Huxley[38] appeared in the British press condemning wildlife "slaughter," and the SPFE pressured the colonial office to halt "inhumane" hunting practices.[39] Alarming declarations such as "hundreds of Natives hunting with poisoned arrows are slaughtering thousands of head of game" were made by groups with vested interests in monopolizing wildlife hunting.[40] European settlers added their criticisms, complaining of lenient sentences for African offenders[41] and once calling for the dismissal of an administrative officer who encouraged Africans to exercise their customary hunting rights.[42] Ultimately, the well-publicized and sensational descriptions of wildlife "slaughter" compelled the secretary of state for the colonies to tell the governor, "I feel sure that you will agree with me that this state of affairs should not be allowed to continue."[43]

The claims in the popular press of wildlife slaughter by Africans were not substantiated in the records and most were immediately refuted by administrative officers. Regarding the above accusation concerning "thousands of head of game," for example, the Member for Agriculture and Natural Resources replied that there were perhaps five animals involved in the incident cited, not "thousands," and perhaps forty Africans, not "hundreds."[44] Of these, he speculated, most were probably porters and noted that the party was small in comparison to European safaris. In a series of responses to the game warden's recommendations to ban poison arrows, several provincial commissioners countered that in their experience death by that method was very quick.[45]

Commentaries such as those cited above reveal that there was more at stake than merely the conservation of wildlife numbers in colonial Africa. Hunting by Africans involved cultural values and practices that offended the sensibilities of Europeans who held fast to their own values and myths concerning wildlife. The enforcement of these values and myths was integral to the legitimation of colonial rule and to reaffirming the superiority of British culture and society. To allow African hunting and forest use to continue would mean placing African culture and resource management practices on equal footing with those of Europeans. The discourse of colonial conservation, then, simultaneously denigrated African land use and natural resource practices and promoted the primacy of European forest and wildlife management techniques.

Supplementing their moral arguments against customary uses of natural resources, the British relied on the ideologies of "scientific" resource management and racial interpretations of African culture. Great weight was given to Professor Troup's assessment of Tanganyika's forestry policies, in which he argued that free issue for Africans hindered scientific forest management. In claiming sole authority to manage the colony's natural resources, resource professionals employed stereotypes of racial inferiority to dismiss Africans' ability to conduct their own affairs. The game warden, for example, argued against the legalization of African hunting, "because the native has not yet reached the stage of civilization at which he is capable of appreciating properly the gifts of nature such as a fine game population and valuable timber forests—and of conserving them."[46] Discussing the possibilities for sustained use management in Tanganyika, a succeeding warden pointed out that "[i]t is well known, however, that in his present state of mental development, the African is incapable of foresight and provision for the morrow."[47] Based on this line of thought, European professionals argued that the control of all natural resources should be placed in the hands of state bureaucracies.

Not all members of the colonial elite, particularly early advocates of indirect rule (Neumann 1997b), were convinced by conservationists' reasoning. Governor Cameron, for instance, envisaged the long-term role of the Forestry Department as advisory only, with Africans performing reserve management. In 1931, the Chief Secretary outlined the government's position for the conservator. The governor felt, wrote the secretary, that "the Government certainly had in mind the desirability of delegating responsibility for the protection of Forest Reserves to Native Administrations."[48] He went on to explain that "the ultimate object would be to effect a complete transfer of responsibility for the actual protection of forests, so that your Department would be in the position of an expert adviser." In his memoirs, Cameron became sentimental about the plight of Africans who had lost legal access to wild meat. Recalling an incident he witnessed in which three people were mauled by a lion, he wrote, "Because these wretched people, with their miserable bows and absurdly tiny shields and spears, having seen the lion strike down an antelope, were endeavoring to take it away from him—for food for themselves. They were afraid to kill a bit of game for themselves, lest they be punished, but were not afraid to try to drive the lion away from the stricken antelope that they might obtain a bit of meat" (Cameron 1939, 239). The governor's compassion for the plight of Africans deprived of hunting rights did not, however, compel him to rescind the colonial game laws.

There were some in the administration who mocked the moral stance of wildlife conservationists and dismissed the professionals' scientific arguments as "lip service" that masked the "real" motivation behind the policies: reservation of the wildlife for elite European hunters.[49] There were costs to reserving wildlife hunting for the pleasure of a privileged few, most of them borne by Africans. "I think a great deal of sentimental twaddle is talked about game by those who confuse the issue," observed an administrator, "which is, simply, whether game is to be preserved for the privileged owner of a .350 Express regardless of the damage it does to the crops or livestock of a man who is at most times on the bare subsistence level."[50] None of this, of course, was lost on the Africans, who were well aware of the links between their loss of rights and European pleasure and profit.

CONSERVATION LAW
AND LANDED MORAL ECONOMY

From the above debates it is clear that the laws severely disrupted African natural-resource use and diminished Africans' control over traditional means of production. However, recovering from the historical record a sense of African responses to these threats to their livelihood is more problematic. Nevertheless, though we have only the jaundiced writings of the game department and other colonial officials, a careful reading of the reports offers evidence for a moral economy within African society that legitimized resistance to the conservation laws of the state. We can detect in the complaints of the rangers and foresters elements of community coherence and complicity among Africans in the face of European assaults on their way of life. Referring to the "slaughter" of wildlife by African hunters, a ranger indignantly observed that "[t]here is practically no attempt made by the people concerned to hide their activities," which their chief, technically a representative of the state, openly supported.[51] When Ikoma hunters threatened game scouts with poisoned arrows should they attempt to stop them from hunting, an official complained that "everybody in the tribe ... knows the names of the members of the tribe involved."[52] Assessing the situation, which continued for years without resolution, the provincial commissioner for the region declared it was uncontrollable because the headmen were in sympathy with the hunters.[53]

Giving due consideration to the source of the accounts,[54] the records show that the Africans were prepared to defend their livelihood against the state's attempts to usurp control. Confrontations between African hunting parties and game rangers—who warned that "the situation will develop into guerrilla warfare"—were not uncommon.[55] The Ikoma, who had a strong tradition of hunting, were particularly defiant and willing to attack rangers attempting to enforce game laws.[56] They openly boasted to the rangers that they would continue to hunt as they pleased and threatened to use poison arrows against anyone trying to stop them. In the forest reserves, meanwhile, guards were "endangering their lives" if they tried to stop traditional burning practices.[57]

Cognizant of the prevailing community acquiescence regarding the open violation of conservation laws, forest and wildlife officials felt communal fines were the solution. The Collective Punishment Ordinance of 1921 allowed the governor to impose fines on an entire village or community if any of the members were involved in harboring criminals. The conservator of forests and others repeatedly requested the government to invoke the ordinance for cases of forest burning.[58] Collective punishments were reasoned to be necessary to halt forest firing because African communities saw it as an acceptable and even desirable practice. As a legislative council member from the Arusha District reasoned, collective punishments would reduce the incidence of fire because "the fire lighter would no longer be a hero but a public nuisance."[59] Weighing the political risks involved in applying the ordinance, the government repeatedly refused. Another proposed solution was to leverage compliance by Machiavellian application of the principle of indirect rule. The game warden Philip Teare, for example, urged "headmen of a village to be held responsible for fire."[60] In the above case of "disorder among the Ikoma," the provincial commissioner withheld half of the chief's salary until such time as the names of the perpetrators were revealed.[61] The record does not show that they ever were, though the chief was eventually returned to full salary.

The discourse of colonial conservation and the discourse framed within an African moral economy were fundamentally incompatible regarding land and resources uses, the meaning of which became a point of struggle. Practices considered by the state to be crimes or privileges were, within the moral economy of African rural life, customary rights and entitlements that came with clan and tribal membership. In the struggle for control, each side could engage their own moral language to legitimize their positions. Though each side might frame the discussion according to their respective visions, both lacked the power to put their claims into practice except within limited areas of time and space. Africans rejected the discourse of conservation and found ways to resist the total loss of access to subsistence resources, but were subject to severe penalties if they were caught. Conversely, during the interwar years, conservationists did not have the political strength to eliminate completely customary rights nor even enforce the legislated prohibitions to which the government agreed.

The Evolution of State Conservation Policy on Mount Meru

On Mount Meru, whose upper slopes contained stands of commercially valuable cedar (*Juniperus procera*) and loliondo (*Olea hochstetteri*), rights to timber and forested lands were the focus of state conservation policy through most of the colonial period. While wildlife was abundant in certain areas of the mountain, and various locally specific game laws were enacted, the fulcrum of struggle between the Meru and the colonial state was forest access rather than hunting. The roots of this struggle reach back to the nineteenth century.

The forest and game reserves implemented by the Germans in the 1890s on Mount Meru mostly remained intact after the British took over the rule of Tanganyika. When colonial authority collapsed during World War I, Meru farmers expanded cultivation and settlement upward, pushing beyond the German reserve boundary at sixteen hundred meters elevation (Spear 1996). In 1921, the whole of Mount Meru, extending downslope as far as the edge of the natural forest (around eighteen hundred meters elevation), was gazetted by the British as the Mount Meru Complete Game Reserve, and in 1928 the old German forest reserve was officially regazetted, its boundaries overlapping the game reserve's. The double designation meant that the area's natural resources were under the strictest state control possible, given the conservation laws existing at that time, and all settlement, cultivation, and hunting was outlawed. It was twice the size of the Arusha and Meru "native reserves" combined.

Because all of the mountain's forests had been under state control since the 1890s, colonial officials assumed that local rights were surrendered long ago. The conservator of forests once explained that the "Germans before proclaiming a Forest Reserve investigated existing rights and generally extinguished them in a proper legal settlement paying compensation."[62] This statement implies important assumptions about the legal procedures followed by the German government, not the least of which is the definition of "existing rights." For example, under German administration, proof of title to land was by authenticated documents only. Legally, "only settlers who could prove grant of land from the German administration, which grants were entered into the register, or those who had documentary evidence of grants from local chiefs or a public authority, had security of title" (James 1971, 14). Under such a highly rationalized system, rights to the commons under customary law were unlikely to be recognized, let alone compensated for. It also implies that the Meru were told the rights they were giving up, understood them, and happily obliged. Shio (1977, 161), for example, pointed out that Meru elders thought they were granting use rights, not ownership. It is likely that they had not turned over their lands as willingly as the conservator implied. Recall that two Lutheran missionaries were killed for being suspected of coveting Meru land just a few years before the Germans gazetted Mount Meru Forest.

Compensation or no, the designation of the Mount Meru forest and game reserves meant a significant alteration in rural production strategies. Some areas of the forest, off-limits to livestock under forest regulations, had historically provided grazing for Meru herds, most critically during the dry season and extended periods of drought. Yet the forest laws allowed some customary practices to continue. Though commercially valuable trees were untouchable, most of the nontimber forest products were still available without a fee. The diversity of plants provided building materials, fuel, medicines, poisons, and uncountable household utensils and tools. In addition, livestock tracks through the reserve, critical for the seasonal movement of the Meru's cattle, were recognized right-of-ways. There was, nonetheless, constant pressure from the Forest Department to eliminate most "privileges."

One such privilege was the keeping of bees and the collection of the honey they produced from the flowers in the thick forest groves.[63] The foresters' wish to eliminate this local use was based on fear that honey gatherers' careless use of fire would destroy valuable timber. They were particularly concerned about this practice on Mount Meru, where beekeeping was an important occupation for some Meru. On the slopes just below the shattered crater there are extensive stands of lush montane forests, where beekeeping was especially productive as a result of the multitude of flowering plants. In this area, four Meru families had exclusive use rights for keeping hives. Each family had its own section of forest. The father passed to his son the right to use the area, as well as the hives themselves. Smoky fires, which had to be carefully watched, were set below the hives to stun the bees while the honey was extracted. Uncontrolled fire or any form of deforestation would obviously devastate honey production, and beekeepers therefore kept strict control over the use of their section of forest. Compensation, to be paid in livestock and *pombe* (local beer), was due the "owners" should someone damage their sites with fire.

Such a system of tenure and labor specialization makes the argument that honey gatherers would be guilty of destroying the forest with fire seem implausible. Nevertheless, honey gathering in government forest reserves was long the number one bugaboo of colonial foresters bent on wildfire control, an almost obsessive preoccupation for the conservator of forests, who called for collective punishments, imprisonment, and even caning for (non-European) fire starters. It was the Meru system described above that was targeted for elimination, for in the Forest

Department's campaign to rid the reserves of honey gatherers, the Meru Forest Reserve was considered "one of the most inflammable" and "should be concentrated on first."[64]

Arguing that honey gatherers are a major cause of forest fires, the conservator of forests eventually requested that it be made an offense to enter a forest reserve for the purpose of collecting honey. When his proposal was reviewed by the Territory's administrators, however, his basic argument—that destructive wildfires and honey gathering were directly linked—gained little support. The local chiefs' description of honey gathering to the commissioner in Mahenge Province, for instance, echoes the system on Mount Meru: "They assure me that honey hunters are most careful of the bee-swarms whose hives they raid for honey. They are well aware that if fire is carelessly employed while raiding the hives the queen bee may be destroyed and the destruction of the swarm will follow. The hunting of honey and the manufacture of wax is regarded as a profession among tribes, and an expert one at that. No novice would be allowed to hunt for honey if he could be prevented. It would be against recognized custom[,] and this is a force strong enough to control such a contingency."[65] The fact that there was ample evidence to suggest that beekeepers were of necessity careful with fire, as well as the questionable political viability of implementing harsh penalties for so ancient a practice, convinced the government that outlawing honey collecting was not a reasonable alternative.

In the face of this opposition, the Forest Department relented somewhat and began to differentiate between beekeepers and honey-hunters—the latter being portrayed as nomadic and irresponsible. After a series of wildfires on Mount Meru in 1937 and 1938, which the conservator of forests blamed on honey-*hunters,* the government was convinced to prohibit entry to the mountain's north side without written permission.[66] A few years later during a period of drought in the Meru area, the conservator again blamed honey-hunters for starting fires and—making an immediate jump from honey-hunters to beekeepers—resurrected the argument for a "total prohibition of the placing of honey barrels in forest reserves."[67] The prohibition was not implemented, but all honey-*hunting* in the Northern Province reserves was outlawed;[68] beekeepers had to obtain licenses and their movements in the Meru Forest Reserve were increasingly restricted.

This example of the struggle over the extinguishment of access rights to one "minor" forest product serves to illustrate the historical process by which rights were slowly eroded rather than eliminated overnight.

This process of tightening state control and whittling away at customary rights on Mount Meru was not confined merely to the forest and game reserves, as the state made broad claims to the ownership of all wildlife and timber in the territory. The Meru area was the test case for the attorney general's 1926 ruling that, since all land except freehold was public land (which included African-occupied lands), the timber that grew upon it belonged to the Crown.[69] In 1926, a correspondence began between the Forest Department and the local administration, which forced a clarification of the department's power to control land use and timber harvesting on African lands.[70] At this time, Mitchell and Kitching were provincial commissioner and district officer respectively for the area that included Mount Meru. In the communications between the administrators and the Forest Department, the right of the department to collect revenue from Meru and Arusha lands was challenged.[71] Seeking Mitchell's support for African control over the trees on their lands, Kitching explained, "It is not however the wish of the Native Authorities that the exploitation of forest produce in the native areas should be left to the caprice of individual natives. There are communal rights which must be safeguarded and there are also the requirements of the district generally which must be met[,] which the Native Authorities recognize and do not resist. The only issue is the form of control."[72] If direct control by the Meru and Arusha is not possible, he argued, at least the timber revenue should go to the Native Treasuries, not the Forest Department.

The matter could not be reconciled locally, and Mitchell asked for a ruling from the Chief Secretary, explaining that "[t]he practice at present is for sawyers to select a tree so situated [i.e., on occupied African lands] and to obtain from the Forest Department a permit to cut it down. They then enter upon the land, fell and cut up the timber, often doing considerable damage to crops in the process, and always being a source of irritation and annoyance to the occupier: royalties on trees so felled are paid to the Forest Department."[73] He asked that the Forest Department not be allowed to issue permits for occupied land. The matter was not confined only to timber rights and revenue, for the conservator of forests was pushing his department's claims even further, reasoning that since "[r]oyalty is chargeable on timber *everywhere*," it follows "that the destruction of forest . . . is unlawful" even on African occupied lands.[74] Based on this logic, the Forest Department tried to order Kitching to stop the Arusha and Meru from clearing trees to make room for cultivation on what all the parties recognized as "native" land.

The Forest Department's arguments on timber rights were couched in the language of "scientific" resource management and conservation, which depicted African use of the land as wasteful and degrading. Where Africans were clearing land on Mount Meru for cultivation, the local forester saw "wanton destruction by natives" resulting from "shifting cultivation."[75] Clearing the land of loliondo constituted a "waste of a valuable commodity," argued the conservator, and it was "incontestable that the heavy timber is not required by the Arusha and Meru natives for their own or their families' sustenance."[76] These arguments—derived from the European-African, scientific-traditional, efficient-wasteful antinomies—obfuscated the fundamental question of *who would control access to and benefits from natural resources.* The Arusha and Meru peoples were well aware of the market value that their trees acquired in the new economy and, believing the trees rightfully belonged to them, wanted the benefits to accrue to their own Native Treasury. The colonial government also desired the revenue for state coffers and, just as important, wanted to keep Africans from competing with European interests in the market economy.

The government eventually responded with a typically ambivalent decision: "[A] native's right in the land is the right of user only; the property in the land remains vested in the Governor. The native's right to the natural produce of the land is limited therefore to the amount required by him for the sustenance of himself and his family[,] and subject to this, Government is entitled to exploit the timber on such land."[77] Yet in the same decision, the conservator of forests was told "to refrain from issuing permits to exploit timber on land occupied by natives until the District Officer has been consulted and concurred."[78] When, some months later, Kitching's successor was asked by Mitchell's successor how the government's decision was working, he replied it was working just fine because no permits had been issued and no cutters had ventured onto African lands. At the end of 1930, without explanation, the conservator himself ordered that no cutting permits be issued "on land claimed by the Wachagga, Waarusha and Wameru."[79] Thus on Mount Meru, as it was with natural resource law and policy throughout the territory, the state made sweeping claims to the ownership and control of resources but found it politically and monetarily too costly to actually implement the claims.

The state's early plans for natural resource conservation and the development of a timber industry, then, necessarily eliminated many of the Meru's local land use practices. Underlying these plans was an ideology of scientific management and nature protection that served to help legitimate the state's claims on control and ownership of all of the colony's resources. Implementation of this ideology was, however, problematic because it required a degree of acceptance and cooperation on the part of the Meru people, who were most negatively affected by the enactment of natural resource laws. The necessary cooperation was not quickly forthcoming.

MERU RESISTANCE TO COLONIAL
NATURAL RESOURCE POLICY

Since the earliest years of colonial rule on Mount Meru, government foresters had difficulty enforcing natural resource laws in the Meru forest and game reserves. The Forest Department was plagued by continuous offenses, involving especially the pressing of customary rights to graze livestock and a constant pressure on the forest boundary. Illegal grazing was the most common offense in the reserve at least since the British period,[80] and various solutions were suggested to slow livestock trespass depending on how the problem was perceived. The assistant conservator of forests in charge of Meru wanted to build a twenty-four kilometer fence along a livestock path to control cattle trespass as they moved through the forest reserve to grazing grounds on the opposite side.[81] When the conservator of forests claimed, however, that cattle were destroying the forest and that "adequate protection of the forests under these conditions is impossible," the district officer accused him of exaggeration and responded with his own exaggerated metaphor: "Waarusha and Wameru country is overcrowded and over-stocked; it is like a boiler with the internal pressure very near the danger mark and when all relief valves should be opened." These tracks were the relief valves: "close them and bursts will take place through alienated land."[82]

The district officer's metaphor powerfully illustrates the political and social tensions on the mountain, which resulted from the extensive alienations for settler estates and forest and game reserves and a hardening of attitudes among government officials. After World War II, technical specialists replaced the sympathetic administrative officers of the 1920s (Feierman 1990; Iliffe 1979). The paternalism of earlier administrators was superseded by the technocratic efficiency of the new breed. This postwar situation in East Africa has been characterized as a "second colonial occupation" (Low and Lonsdale 1976, 12) marked by a deeper penetration of the colonial state into nearly all aspects of rural African society (Beinart 1984; Feierman 1990). The change had a chilling effect on the already strained relations between natural resource officers and Africans. A letter to the Member for Agriculture and Natural Resources described the problem with the new crop of foresters: "They at times forget that in creating a forest reserve they generally are eliminating some ancient rights and are giving the individual very little in return . . . The result of all this is that there has been a steady hardening of African opinion against all forest reserves[,] and this attitude is probably the greatest single adverse factor in the creation of an adequate forest reserve in Tanganyika."[83] Given the land crisis for the Meru, it is not surprising that perhaps the most significant problem confronting these forest bureaucrats was the encroachment of settlements and cultivated land into the reserve.

In order to be successful, an encroachment had to go unnoticed by the reserve officials, and apparently the practice approached an art form among Meru peasants. In an almost admiring account of a "most deliberate and premeditated encroachment into the forest reserve," a government surveyor described how local residents carefully repositioned the concrete beacons thirty to forty-five meters parallel to the legal forest boundary.[84] They dug guide trenches, erected guideposts and built cairns with a beacon in the center as an "exact replica of the original." "To fully comprehend this master-piece of encroachment is difficult," the bewildered surveyor wrote, "but I feel that this sort of thing has gone on quite often on both Meru and Kilimanjaro over the past thirty years."

Even if the perpetrators of encroachment and livestock trespass were caught by forest authorities, getting the Meru courts to prosecute their cases was another matter. In pressing the prosecution of forest violations, department officials could find themselves drawn into a frustrating game of cat and mouse with the Meru authorities. Correspondence in 1948 between the forester in charge of the Meru Forest Reserve and the Arusha district commissioner is illustrative of the situation.

The forester was complaining about the lack of cooperation on the part of Meru authorities in enforcing forest laws, specifically a case of illegal timber cutting that Jumbe Ndamu failed to act upon and another case in which a forest guard caught a Meru resident with twenty-eight cattle and eighteen goats inside the reserve, but which was later dismissed in a local court.[85] After interviewing the principals involved, the district commissioner replied to the forester that the *jumbe* denied he was doing nothing and said that he was instead only waiting for the forest guard to give him the name of the offender.[86] In regard to the grazing case, the district commissioner said that in Mangi Sante's court, people were paraded before the forest guard who failed to identify the accused or his sons. Mangi Sante, in fact, turned the indictment around, and the district commissioner reported to the forester that the chief was highly suspicious of "your two Pare Forest Guards," whom he believed were "condoning offenses in the reserve." Sante suggested that the Forest Department hire Meru guards who could identify the cattle and their herders. The correspondence continued, with the forester contradicting Mangi Sante, complaining that his forest guard never saw any "identity parade" in court.[87]

And so it went on Mount Meru throughout the colonial period. As the forester pointed out in response to Sante, hiring Meru guards for the reserve had "been tried, but with no success in Meru country."[88] The forester could not rely on their testimony in court, let alone rest assured that they were not giving the forest away to their friends and family. When the forester found a budding settlement of three houses inside the reserve, the local elders told him that they thought it was legal because the Meru forest guard had personally shifted the boundary.[89] Previous to the discovery of this particular encroachment, the Meru guard had been sacked by the Forest Department for giving out agricultural plots in the reserve.

As late as 1960, cases of encroachment brought before the Meru courts were still being dismissed. In one case, four people were arrested for encroaching and taken to the court in Nkoaranga. The *hakimu* (Meru Native Authority judge) told the forester they had been fined, but would appeal. A month later when the ranger asked about the appeal, the *hakimu* said he had already acquitted them, offering no explanation for his decision.[90]

Eventually it became clear to government authorities that it was nearly impossible to prosecute forest offenders in the lower Meru Native Courts. A memo from the assistant forest conservator concerning a Meru judge in the Native Courts sums up the situation. He wrote, "The above Judge is repeatedly refusing to deal with Forest Offenders when cases of encroachment or illegal cultivation and residence are brought in by my staff; instead of convicting the offenders, he gives them a certain period within which he allows the offenders to move out of their farms/encroachments. In one case he has done so contrary to a court order by the Resident Magistrate, Arusha."[91] Following this, the district commissioner directed that all forest encroachment cases should be heard at Baraza A (the central Meru Native Court) in Poli, while cases on other violations, such as grazing trespass, would still be heard in the lower courts.[92]

Since the arrival of the Germans, the pattern of natural resource management and access control on Mount Meru (and in Tanzania in general) has been one of increasing state intervention and a steady erosion of the Meru's customary rights. Under the independent government, local control has been eroded even further. In 1963, when the TANU government abolished customary political authority, it eliminated most remaining local control. "Native forest reserves" in Meru, set up by the British to be managed locally with revenue going directly to the Meru Native Treasury, were turned over to (non-Meru) TANU party officials, and the revenue redirected to the district treasury. Rights of access to the government forest were gradually whittled away until all access to the "natural areas" of the Meru Forest Reserve was completely eliminated in 1984 by order of the director of forestry and beekeeping.

Elucidation of this historical pattern is of critical importance for understanding contemporary conflict. Colonial authorities were well aware that the new natural resource laws were obstructing African practices and eliminating what were perceived as inalienable rights, but proceeded in the name of efficient and scientific management. The implementation of game and forest laws, by unilaterally depriving the Meru (and other peoples) of customary access to land and resources, spawned a subculture of resistance (Scott 1985) to their enforcement. The manner in which this subculture operates is perhaps best exemplified by the actions of the individuals who were servants of the Crown yet simultaneously deeply tied to the moral economy of the village—the Meru judges, headmen, and forest guards. The dismissal of cases of forest trespass and the incidents of guards looking the other way as villagers encroached on the reserve are examples of community resistance to state policies that threatened established production and reproduction strategies. If the accounts of the foresters and administrators reveal anything about the history of natural resource crimes in the Meru Forest Reserve, it is that their perpetrators cannot be characterized as independent-acting, asocial criminals. Forest encroachments were well planned and organized tactical maneuvers to reappropriate land from the reserve without the management being any the wiser, and their success necessarily required a degree of community cooperation. When individuals *were* caught grazing or encroaching, the local communities, including their courts and elders, were actively complicit in seeing that they were never punished. We can see in this solidarity of opposition to forest laws, strong parallels with the organization of resistance to the eviction from Ngare Nanyuki.

This historical examination also reveals that, far from being a homogeneous entity, the colonial state was strongly divided over the conflict between African rights and natural resource management and conservation. Those administrators who believed that African rights had primacy over the designs and policies of colonial resource professionals found themselves at odds with politically powerful conservation advocates in London. The interests of British conservationists in African wildlife were represented principally by the SPFE, which, as will become evident in the following chapter, played an instrumental role in the design of Tanzanian wildlife policy and in efforts to create the first national park in a British-ruled African territory. For the SPFE and the British classes that they represented, nature protection in Africa was a deeply moral issue, and the symbolic meaning attached to national parks equally as important as their ecological significance.

4

Protecting the Fauna
of the Empire

The Evolution of National Parks in Tanzania

As disruptive as hunting and forest laws were to customary land and resource practices, national parks went farther by totally
eliminating all rights. Where the game and forest conservation laws
excluded certain customary uses, the national parks excluded people
from the landscape entirely. Those who were displaced in the process
watched European interests in wildlife triumph over their customary
patterns of land use. In a 1957 memorandum, Maasai residents in
Serengeti observed that "[f]rom time to time we see [white] hunters
passing . . . with the trophies of the animals that they have shot . . . It is
these same people and their friends who wish to evict us from the
National Park[,] yet we think it is they who are the enemies of game
rather than us."[1] In the dominant historical narrative that followed, the
African farms, pastures, and settlements that spread over the land prior
to its protection in national parks disappeared and were replaced by
vacant "wildernesses." In actuality, the establishment of national parks
in colonial Africa often involved the state denying a disenfranchised
society of peasants and pastoralists access to traditional resources, dislo-
cating land use practices and entire settlements, and thereby threatening the communities' very existence (see, e.g., Turton 1987; Arhem
1984; 1985; Diehl 1985; Kjekshus 1977; Marks 1984; Ranger 1989).

This chapter examines the evolution of Tanzania's national park system from the interwar period to the present, paying particular attention
to the role of international conservation organizations and local
responses to the loss of land and resources. One of the main purposes

of the chapter is to explore the relationship between the colonial state structures of nature protection and present-day institutions. Unfolding more or less chronologically, it describes the political turmoil associated with the establishment of what was intended to be the first national park in British-ruled Africa, Serengeti, and the enthusiastic adoption of the national park ideal by the independent government. I strive to demonstrate that the bureaucracies, laws, reserve boundaries, and ideology of conservation have their origins in the era of European colonialism, and to reveal the ways in which these were transferred to the leadership and bureaucracies of the postcolonial state.

While the possibility of economic gains through tourism helped to motivate the state's efforts to protect and control natural resources, national parks are a unique form of resource management. They served as powerful symbols of a European-based vision of what Africa should be, a landscape encompassing socially constituted notions of culture, nature, and civilization. Thus, in colonial Tanzania the central issue in the establishment of Serengeti was the question of human presence in the national park ideal. The question was answered in Tanzania during the colonial period, producing in the process one of the bitterest and most protracted natural resource conflicts in the country and greatly influencing the tenor of wildlife conservation for decades afterward.

Protecting the Fauna of the Empire

In 1928, not long after he was appointed, Tanganyika's first director of game preservation recommended that Mount Meru, along with Ngorongoro Crater and Kilimanjaro, be designated as a national park, though no such designation yet existed in the laws of British-ruled Africa.[2] The most forceful advocacy for creating parks in Tanganyika, however, came not from colonial resource professionals but from politically powerful conservation societies in England. Dated a month after the game warden offered his suggestion, a confidential letter sent from the secretary of state to Governor Cameron provides the first official documentation of interest in creating an international agreement to protect African wildlife.[3] Under the auspices of the Society for the Preservation of the Fauna of the Empire (SPFE), the governor was informed, Captain K. F. Caldwell would develop a proposal for obtaining such an agreement. The matter rested there for five years, when the secretary wrote to Governor Stewart Symes that the 1931 Paris Congress for the Protection of Nature renewed interest in an international agreement.[4] While thirteen European countries were represented at the congress, the delegation from England, led by the SPFE, saw themselves as the main hope for organizing definitive action.[5]

In conjunction with the push for an international agreement, the SPFE sent a representative, Major Richard Hingston, to Tanganyika to investigate the need and potential for developing a nature protection program. With the full endorsement of the Colonial Office, Hingston was allowed access to and the assistance of Tanganyikan officials.[6] The report that resulted contains interesting conclusions regarding national parks in the colony. It stresses that "[t]he keystone of the report is the recommendation regarding the formation of a number of national parks without delay."[7] Major Hingston specifies three areas, Serengeti, Kilimanjaro, and the Selous Game Reserve, as having the qualities of a national park. (The Tanganyika game warden later recommended that Lake Rukwa and "the great Mount Meru Reserve" be added to the list of potential national parks.)[8] His recommendation for Serengeti was a staggering 25,900-square-kilometer area (nearly the size of Belgium) stretching from the shore of Lake Victoria eastward beyond Lake Natron. Within the park envisaged, the Mwanza District alone contained seventy thousand Africans whose lives would be under the control of park authorities.

Hingston's report fueled the SPFE's efforts to forge an international agreement, and the Society's president, Lord Onslow, cited it in his address to the 1931 Paris Congress.[9] Equipped with Hingston's report, the SPFE took on the task of chairing the preparatory committee to draft a set of proposals for an international agreement. The committee put great emphasis on national parks, "as put into practice in Canada, the United States, South Africa and the Belgian Congo," as the best strategy for preserving wildlife.[10] They proposed to include in the agreement an obligation on the signatory powers to "explore the present possibilities of creating National Parks."[11]

The proposal set up the arguments for establishing national parks and is therefore a useful document for examining stated rationale and for gaining insight into the logic behind the proposed park model. The Society's rationale can be broadly categorized as ecological, material, and moral. First, the parks were intended to play a role as sanctuaries for the declining wildlife populations. Human population growth and overhunting were cited as the major threats, and it was argued that only parks and reserves could save some species from extinction. Second, they argued that keeping wild animals confined to parks would "be highly advantageous to the material interests of the native population of Africa" because it would reduce problems such as crop and livestock raiding. Tourism, of course, was also seen as a direct material benefit to be derived from the new parks. Finally, there were appeals to a higher cause—for preserving nature for nature's sake—and references to the moral obligations of the colonial governments. In regard to nature protection, the Society claimed, "[e]very government has therefore, a responsibility in what it may do or leave undone, to the educated opinion of the civilized world."[12]

More significant, both from a theoretical standpoint and for understanding the historical basis of contemporary social conflict in the national parks, are the Society's suggestions about the relationship between parks and people. In casting about for models, they logically made reference to Yellowstone National Park, the first such protected area established, and Kruger National Park in South Africa, which was overseen by an independent board of trustees.[13] Significantly, both the U.S. and South African models prohibited any human settlement or activities other than tourism. Nevertheless, the politics of establishing parks in the colonies (particularly in Tanganyika, as will become clear below) inhibited the Society from taking a definitive stand on this issue. They made clear that humans were not welcome in parks, but allowed that under the prevailing political climate preservation "cannot be pushed to a point at which it seriously conflicts with the material happiness and well-being of the native population."[14] Reflecting a popular racial stereotype of "primitive" Africans as part of the natural landscape, they pointed out that sometimes a "native" presence in the parks may be tolerated. Such was the case in Parc National Albert, where "the Pygmies are rightly regarded as part of the fauna, and they are therefore left undisturbed." Their perspective on the question of society-nature relations in national parks spurred constant political disputes, and its ambivalence would not be resolved for another twenty-five years.

The second draft of the "Report of the Preparatory Committee" was subsequently circulated by the secretary of state for the colonies to African colonial governments for comment.[15] Far and away the most prevalent theme of the comments of Tanganyikan officials was a concern that the parks would interfere with local customary rights.[16] Specifically mentioned were rights to grazing, hunting, and minor forest products. Consequently, Acting Governor D. J. Jardine replied to the secretary that a clause be added "to the effect that the protection of vegetation in national parks does not interfere with the rights at present enjoyed by the native inhabitants to pasture or to forest produce."[17] The secretary's eventual reply points out that nothing in the national park definition makes "native" habitation inconsistent, "provided that [native activities] are controlled by Park authorities."[18]

The Convention for the Protection of the Flora and Fauna of Africa was held in London in 1933, resulting in an international agreement that closely followed the SPFE's Preparatory Committee's proposals. The definition of a national park, however, was "substantially amended," although the initial ambiguity over the rights of people remained. A summation of the London Convention circulated to the colonies noted that "National Parks need not necessarily be entirely devoid of human habitation, although human settlements should naturally be avoided as far as possible," and stressed again that any human activities would be controlled by park authorities.[19] Despite, or perhaps because of, the haziness surrounding what was politically the most explosive aspect of national park designation, the agreement was signed by the colonial governments concerned, including Tanganyika. It should be borne in mind that signing was made even more painless by the fact that the Convention only obligated the signatories to look into the potential for establishing parks.

The Tanganyika government seemingly took the agreement to heart. In November 1937, Governor Harold MacMichael sent a confidential dispatch to the Colonial Office, informing the secretary that the government was preparing to demarcate a 7,800– to 10,400–square-kilometer area out of the Serengeti Closed and Complete Game Reserves and Ngorongoro Complete Game Reserve as Tanganyika's first national park.[20] Close scrutiny of the historical records, however, reveals the colonial administrators to be reluctant conservationists at best. Ten months before Governor MacMichael's dispatch, Tanganyika officials were complaining of England's Lord de la Warr's interest in creating a park at Serengeti. "We have resisted all attempts to create a national park or adhere to any international convention relating to the preservation of fauna," an administrator wrote. "The pressure from home may, however, be too much."[21]

As the pressure from conservation societies and individual conservationists grew, the lines of conflict within the colonial government and British society in general came into sharp relief. On one side were the officers responsible for administering the territory, one of whom referred derogatorily to conservation societies as "a pest."[22] They were, by and large, of the opinion that conservationists' strategies conflicted with African rights and therefore threatened political stability in the territory. Fearful of "letting fanatics loose in these matters,"[23] they were loath to see the foresters and game wardens gaining power in the territory with the support of conservationists in the metropole.

Throughout the period of the British Mandate, administrative officers continually criticized the national park proposals much as they did the game and forest laws. From their perspective, for example, hunting by African residents inside the proposed parks was acceptable since, "when such natives have enjoyed customary rights of hunting[,] there is no reason or justification for depriving them of these rights."[24] A. E. Kitching, a district officer and later provincial commissioner, was particularly vocal in his opposition to the national park proposals: "The Hingston recommendations . . . pay no regard to native interests. They involve the alienation in perpetuity of thousands of square miles of the land of the Territory . . . to 'create the finest nature park in the Empire.' The recommendations appear to me to be so wrong in principle as to make any detailed examination unnecessary."[25]

Officers such as Kitching, however outspoken, were no match for the political heavyweights in the metropole, led by the SPFE, behind the drive for national parks. The SPFE originated as a small delegation formed to lobby its members' peers and relatives in the colonial bureaucracies to protect aristocratic interests in African hunting (see Fitter and Scott 1978; MacKenzie 1988). Edward North Buxton, Verderer of Epping Forest and a hunter of African big game, initiated the organization after successfully lobbying the Colonial Office in 1903 to halt the decommissioning of a game reserve in Sudan. The majority of the original core membership consisted of titled English landowners, a pattern that continued through the interwar years.[26] Many members thus had ready access to the highest levels of government through their own extensive social and political networks.[27] As members of Parliament, they were able to call government officials to the House of Lords for questioning and lobbying.[28] As well-connected members of the aristocracy, often with past or current appointments in colonial service, they were readily received in delegation by successive secretaries of state for the colonies.[29] In this way they were able to bring their interests directly to the attention of the secretary of state, which then "could be communicated to the Governors" (SPFE 1930b, 14).

Richard Onslow, Fifth Earl of Onslow, was president of the SPFE in the critical decade of the 1930s when the idea of creating national parks in colonial Africa was gaining momentum. Under his guidance from 1926 to 1945, the SPFE was instrumental in formulating the national park ideal in Africa (see Neumann 1996) and took the lead in pushing for parks' establishment. While president of the SPFE, he was also a member of the House of Lords, holding the powerful offices of deputy speaker of the House and chairman of committees from 1931 to 1944. Through his family, he had numerous ties to government and the Colonial Office. Onslow's father (the fourth earl), for example, served as governor of New Zealand (1889–92) and undersecretary of state for India under Balfour (1902–5). Lord Onslow's sister married Lord Halifax, viceroy of India from 1926 to 1931 (when he was Lord Irwin) and foreign secretary in 1938. Before becoming chairman of committees, he held a series of minor government appointments, at one point representing the Colonial Office in the House of Lords.[30] Such family connections and positions allowed SPFE members entry to the highest levels of the British government and the Colonial Office,[31] where they could use their considerable influence to, in some cases, virtually write colonial park and wildlife policies (Neumann 1996).

Through such politically powerful individuals as Lord Onslow, the idea of the superiority of European natural resource practices and elite conceptualizations of nature and culture could be translated into the concrete realities of conservation laws and national park boundaries. The British were attempting to implement a mythical vision of Africa as an unspoiled wilderness, where nature existed undisturbed by destructive human intervention. It was necessary to dehumanize the Africans who lived and worked in this virgin landscape so that reality would fit within the vision. "Primitive" Africans were often simply regarded as fauna, particularly by the class of military officers from which the colonial resource institutions initially drew their personnel.[32] The possibility of protecting them along with the wildlife could therefore be given serious consideration. It was the Europeans' prerogative, moreover, to determine the character of primitive culture. That is, just as there was a particular European conception of unspoiled nature that Africa represented, there was an interrelated concept of primitive human society. Both had more to do with European myths and desires than reality. As will be seen below, those Africans whose behavior did not fit with British preconceptions of "primitive man" could not be allowed to remain in the national parks (*the* symbol of primeval Africa), regardless of their claims to customary land rights. National parks were at once symbolic representations of the European vision of Africa and a demonstration of the colonial state's power to control access to land and natural resources. The natural resources of the colony were claimed for the Crown, hence nature protection and the legitimation of political claims were closely linked. This linkage was symbolically demonstrated by such means as recounting the moral example of the king and queen's hunting trip to East Africa (Neumann 1996) and the coordination of Serengeti National Park's christening with the queen's coronation celebrations.[33]

The Creation of Serengeti National Park

Despite the governor's 1937 dispatch on the planned establishment of Serengeti National Park, the colonies were not moving fast enough for the government in London. A circular from the secretary of state sent to the African colonial governments urged progress in preservation in time for a 1939 international conference on nature protection, so that the United Kingdom could show "a position more in line with the requirements of the [1933] Convention."[34] To this effect the Tanganyika government stepped up action on the Game Ordinance for the Territory, which had been pending for over six years and contained a clause that would declare Serengeti the first national park in British colonial Africa.

A special committee was established by the governor to review the bill, and it reiterated government concerns over indigenous rights in the proposed park. The committee recommended "that the requirements of the National Park not be allowed to interfere with existing grazing or water rights."[35] When the draft ordinance circulated among the administrative officers, Kitching, now a provincial commissioner, was irate over the legislative process and again concerned about the infringements on African rights. Kitching insinuated that the park proposal was being ramrodded through: "The Bill . . . has not been seen by the Provincial Commissioners since July 1934, when the first draft was circulated . . . The District Officers have had no opportunity of commenting upon the Bill or of explaining its provisions to the Native Authorities and ascertaining their views, although it touches matters of great interest to the native people."[36] He asked for a full investigation of customary rights within the proposed boundaries and that the clause that was to establish the park be deleted from the ordinance so that it could be fully debated in the legislature. His requests were ignored and the Game Ordinance was passed by the Legislative Council in May 1940.

Serengeti was now a national park in name, but it fell substantially short of what the SPFE had envisaged. There was no designated management staff nor any special funding for the park. The issue of human settlement and the control of human activities by the government had not been resolved. The difficult task of documenting indigenous rights had not been undertaken. Most important in regard to compliance with the 1933 Convention, the park lacked an independent public body to oversee it. Consequently the pressure on Tanganyika from the Colonial Office, spurred on by the SPFE, continued. In 1946, the Society again prodded the Colonial Office in London, sending a deputation to complain that compliance with the Convention was not being met.[37] The London office passed the complaints on to the colonial governors, "suggesting" that they organize an interterritorial conference to deal with the issues.[38] The Tanganyika government almost immediately began working on the necessary legislation.[39]

Ultimately, Serengeti National Park took on great symbolic significance for conservationists, and this fact helps to explain the SPFE's unceasing lobbying efforts. It came to embody for the Society much of what it wished to accomplish for wildlife preservation in Africa. The Society stressed to the government in London "that the first National Park to be established in a Mandated Territory of British Africa should be modeled as closely as possible on the provisions of the Convention." The Serengeti National Park was likely to serve as a model for others in the British African colonies and elsewhere, and it is on this account that the committee addressed the secretary of state while its constitution was as yet undetermined.[40]

The Society's efforts were focused on establishing a national park in the Northern Province at Serengeti according to their agenda and their own hurried time frame, without ever resolving the key issue of human presence within the boundary.

In commenting on the report from the 1947 interterritorial conference, the secretary of state for the colonies took the opportunity to remind Tanganyika's governor about the colony's obligations regarding the 1933 London Convention. Despite Tanganyika's intention to implement a special National Parks Ordinance, the secretary wrote that the colony had not done enough to comply with the Convention. He stressed to the governor that the Convention's resolution regarding "the creation of National Parks . . . is from the long term point of view the most important," and urged the colony to look for "other areas suitable for constitution as National Parks." Bearing in mind the ceaseless political pressure brought to bear on the colony by the home office, Tanganyika's headlong rush into political disaster in Serengeti is easier to comprehend.

INCREASING RESTRICTIONS, GROWING UNREST

The Legislative Council passed the National Parks Ordinance in 1948, which established an independent Board of Trustees and designated Serengeti as the first park under the new law. However ambivalent the conservationists may have been over the question of human habitation, the law was not. Significantly, the ordinance did not outlaw human occupation and in fact explicitly permitted the unhindered passage of people "whose place of birth or ordinary residence is within the park."[41] Still, this left open the difficult question of who did or did not have a legal right to be in the park, which, as will be seen, was political dynamite and almost impossible to determine given European knowledge of African history and culture during colonial rule. The proposed park boundaries were immediately disputed by Africans living nearby, and they were not finalized until 1951.[42] Though the secretary of the new Serengeti National Park Board of Trustees reassured the government that "the rights of the Masai . . . to occupy and graze stock in the Park are unaffected by the Ordinance,"[43] less than a week later the new park warden wrote that the trading post and Maasai cattle market in Ngorongoro must be removed as they "interfered with the amenities of the park."[44] It soon became clear that the ordinance constituted a formula for unrest.

What did the Africans—the Maasai, Ndorobo, Ikoma, and Sukuma peoples—who used or lived in the proposed park have to say in regard to the loss of their rights? Not surprisingly the government records are not overflowing with African points of view on this issue, though we do know there were "constant and vexatious clashes of interest."[45] One of the more intriguing documents from this period is a detailed statement prepared in the name of the "Masai of the national park," outlining their historic land claims, grazing and water tenure system, and subsequent disputes with the park administration. A passage from this document declares,

We do not know precisely how long Masai have been in occupation of the country in the neighbourhood of Ngorongoro Crater but the period exceeds 150 years. The present Laigwenan of Ngorongoro was born in the Lerai forest in the Crater during the first decade of this century[,] as was his great-grandfather. Other elders are alive today who are able to testify similarly that their ancestors lived in the Crater Highlands before the circumcision of the Il Merishari [c. 1810] . . . The Lerai forest has been a sacred grove for the whole Masai tribe for many years, being used for rainmaking and fertility ceremonies. In this forest the graves of important Masai elders who died several generations ago can even now be identified by their descendants. (Today we are forbidden to enter this forest and this grieves us sorely.)[46]

These ancestral claims were summarily dismissed by international conservationists who wrote, "The Ngorongoro Crater was not original Masai land, nor was much of the contiguous mileage."[47]

As might be imagined, some of the tensest confrontations involved Africans exercising their customary hunting rights and game officers trying to stop them. Field officers frequently complained of having hunting weapons turned against them. One game patrol in the Sukuma section of the national park was castigated before the chief's *baraza* ("native" court) for arresting locals. "I have called you here," the chief was quoted, "to tell you before the *baraza* that your cases will not be dealt with, and that we do not care of [*sic*] what you call the game reserve here. You must know that we are not prevented to kill game in what you call the game reserve. Now all the offenders you brought before me the other day will not be called to the *baraza*. Go away now, I do not want to see you in my place."[48]

Pariah-like, a game scout was unable to carry out his duties within the moral economy of rural communities, for "should he foolishly prosecute some of the locals he becomes 'public enemy number one' and cannot get food, beer, or bed."[49] African hunters in this area were, however, highly valued by the communities that they helped to support. A standard practice among societies with a strong hunting tradition was for the hunter to take the animals' hindquarters and give the remainder to the local village.[50] "Poachers" were maintaining the expectations and obligations of the moral economy; game scouts were violating them.

In another part of the park, the Maasai were also resisting the criminalization of their customary practices. In the early 1950s, a special administrative post had to be set up in Ngorongoro Crater because the "Masai were openly defying the Park laws, and the political situation had consequently become explosive and a magnet for agitators."[51] Deprived of a political voice the Maasai found ways to protest their predicament, starting fires "with malicious intent."[52] Given a chance to speak out about their situation, the Maasai wondered why they were not allowed to represent themselves in Dar es Salaam and why their rights were being restricted because of the decline in wildlife populations, which, they argued, was the result of hunting by Europeans.[53] "We find it difficult to understand the attitude of some Europeans," read one Maasai memorandum, "who at the same time that they cry out for the protection of game[,] take out licenses to hunt and kill elephant, lion, and other animals."[54]

The Maasai and other displaced peoples were fully cognizant to the fact that their rights were being sacrificed for the interests of privileged outsiders. They had initially been evicted from the Ngorongoro Crater section of the national park under the German colonial administration. The Germans had alienated the crater lands for a settler estate owned by the Siedenkopf brothers, who farmed and raised livestock there. The pastoralists, who had carefully avoided contact with wildebeest during its calving period because of a contagious animal infection, watched the new settlers attempt to wipe out the threat. A Maasai memorandum recalled: "The German, Siedenkopf, who used to rear stock in the Crater killed hundreds of wildebeest in an attempt to prevent his cattle from being infected with malignant catarrh. We were not happy to see this slaughter and asked him to stop."[55] The slaughter stopped when the farm was confiscated after World War I by the British government, which converted it to a game reserve. After the reserve became a national park, it was the European hunter-naturalists and game and park officials who threatened a new round of evictions.

One of the government's responses to the unrest in the 1950s was to write down in a "bill of rights" precisely what the Maasai who were living in the park could expect. In keeping with the top-down orientation of national park conservation in colonial Tanganyika, the Maasai did not participate in drawing up the "bill of rights" and were only allowed to see it after it had been completed.[56] In this way the conservation professionals kept within the written law but violated its spirit. That is, the National Parks Ordinance explicitly protected customary rights, but it was the *park officials* who decided what those rights were and who would be entitled to them. Park administrators determined the extent

of the customary land of the Serengeti Maasai, generally ignoring seasonal and longer-term livestock movements and dismissing much of
their land claims. The Serengeti Maasai made a formal claim to "[t]he
right to move without restriction within the boundaries of Masailand.
This means that a Kissongo from Kibaya may settle in Loliondo if he
wishes, and a Purko from Loliondo may settle in Kibaya. We have always
practiced interchangeability of grazing and watering facilities with all
that this implies. We feel that this was not taken into account when the
National Park was first established, it then being assumed that only
those Masai who were actually born or resident within the Park at that
time had any rights within the area."[57] In the end, the park regulations
restricted their movements to such a small area that the provincial commissioner was moved to call their decision "a complete breach of faith
with the Masai."[58]

Fulfillment of the European vision of primitive Africans living "ami-
cably amongst the game"[59] meant freezing economic development and
cultural change within resident communities. Under their "bill of
rights" the weapons that the Serengeti Maasai were allowed to carry
were restricted to "spears, swords, clubs, bows and arrows."[60] Those
Maasai that were allowed to stay were placed under strict control to
assure that they remained "primitive." In defining homestead building
codes in the Maasai "bill of rights," the chairman of the National Park
Board of Trustees "explained the reasons why they wished the word
'traditional' to be inserted in the draft definition in order that the Masai
living in the Park should retain their present primitive status. He and
the D.N.P. felt that if the Masai changed their habits and wished to
build other types of housing, they should do so outside the area of the
Park[,] which is to be reserved as a natural habitat both for game and
human beings in their primitive state."[61] In the mind of the conservationist, the Maasai in the park were a colonial possession and could be
preserved "as part of our fauna."[62]

The records provide no evidence of cooperation on the issue of
rights or that the regulations set down were negotiated settlements.
Quite the contrary, the Africans who were affected most were probably
the last to know. In asking for an alteration of the park's boundaries in
order to recover lost tribal lands, a colonial officer commented that the
Sukuma "were arbitrarily deprived of it years ago, without any prior
consultation with their local representatives."[63] It can be argued that
the entire process of identifying rights was beyond the capabilities (and
desires) of the government, as the authorities were well aware.[64]

SETTLING THE QUESTION OF HUMAN RIGHTS
IN THE NATIONAL PARKS

To grasp the magnitude of international conservationists' influence and their ability to push the human rights issue in Serengeti to its denouement, we need to step back and view the conflict within its historical and regional context. In postwar East Africa the two dominant and inextricably linked features on the political landscape were African nationalism and land rights. In 1951, the same year that Serengeti's boundaries were finally set, the Mau Mau Emergency was declared in Kenya, just three hours' drive from the park. The fighting there between the African Land and Freedom Armies and the colonial government continued throughout the period of greatest unrest in Serengeti. In 1951, in another part of the same province as Serengeti, the Tanganyika government forcefully evicted and destroyed the property of hundreds of Meru peasant farmers to make way for a dubious European settler scheme. The Meru were secretly in contact with nationalist leaders in Nairobi who helped them organize their resistance to the eviction.

In Serengeti, meanwhile, boundaries were finalized and the conservation professionals began making plans to evict cultivators living within the park.[65] Acknowledging that the cultivators were legally protected by the National Parks Ordinance, officials proceeded anyway, though worrying that "their eviction is not going to pass unnoticed among local agitators and it is therefore important that it should receive legal sanction."[66] The provincial commissioner warned that the "[g]overnment could not tolerate any forced eviction, particularly with the Meru problem hanging over our heads."[67] In fact, the Serengeti evictions had previously attracted the interest of the Kilimanjaro Citizens Union, which had pledged to support the evicted families in their fight.[68] As was detailed in chapter 2, the Kilimanjaro Citizens Union had close ties to the Kenya African Union, and Tanganyika authorities feared that their subjects, including Maasai, would be "infected" with Mau Mau through such alliances. In the Northern Province, the government was concerned about "visits of Kikuyu/Masai half-breeds who are . . . possibly smuggling illicitly gained arms and ammunition into Kenya."[69] In the "general state of unrest" resulting from Mau Mau, the Meru eviction, and the troubles at Serengeti, the Maasai, it was observed, had "gone in for widespread cattle raids and the moran have got out of hand."[70] Government intelligence reported meetings between Maasai and "Kikuyu Mau Mau" in Loliondo on the eastern border of the park.[71] Eventually, the government attempted to avoid further fueling of nationalist flames by convincing the Maasai Native Authority to actually give notice of eviction to the Ndorobo who were staying inside the park on lands the government considered to be traditionally Maasailand.[72]

As East Africa threatened to explode over the issue of African land rights, the conservationists were advocating, and gaining government support for, the dislocation of hundreds of families to create what was popularly viewed as a playground for white tourists. Rather than retreating in the face of growing nationalist sentiments surrounding African land rights, conservationists pushed the issue. The difficult questions concerning human rights that had been avoided during the campaign to establish a national park could no longer be neglected. As unrest among park residents grew, the conservationists' position hardened and became less ambiguous over the issue of human occupation in a national park. "The interests of fauna and flora must come first," a park manager wrote, "those of man and belongings being of secondary importance. Humans and a National Park can not exist together."[73] The protests and discontent among the indigenous residents and the growing intransigence of conservationists threatened to bring their vision crashing down. Governor Twining had earlier warned the Serengeti National Park Board of Trustees, "If the administration of the National Park were likely to cause any serious threat to the maintenance of law and order, or to the implementation of the Government's policy in respect of the African population generally, then I should not hesitate to introduce into the Legislative Council the measure necessary to rescind the proclamation whereby the Serengeti National Park was declared."[74] Twining of course saw the Serengeti evictions within the larger landscape of political violence erupting across East Africa, which he was struggling to minimize in Tanganyika.[75]

The issue of human rights ultimately had to be resolved, however. Conservationists' original basis for accepting a human presence within the park, collateral with political expediency, was their belief that the Maasai would not detrimentally affect nature preservation efforts. Like the "Pygmies" of Parc National Albert, the Maasai were imagined to be living more or less harmoniously with nature because they were nomadic, did not hunt, and generally did not cultivate. When Africans did not live up to European stereotypes, attempts were made to make them conform and, in the context of Serengeti, these attempts generated more conflict. For instance, some Maasai did in fact cultivate,[76] though the national parks director tried to explain the presence of cultivators in Ngorongoro Crater as a result of the Maasai having become "much adulterated with extra-tribal blood."[77] Based on British interpretations of African culture, the legal logic by which these cultivators could be evicted ran as follows: since we know that Maasai do not cultivate, any cultivators in the crater must be non-Maasai, and since no non-Maasai may live in Maasailand without a permit from the Native Authority, they are therefore without legal rights.[78]

The Maasai were not the only residents in the park. The Ndorobo and the Sukuma, among others, cultivated the land and hunted wild animals. Though nothing in the law prohibited the practice, park officials had all along planned to evict cultivators, and soon after the park was gazetted they began plans to amend the ordinance to explicitly forbid cultivation.[79] The 1948 National Parks Ordinance contained an inherent contradiction. The "saving clause" read: "Nothing in this Ordinance contained shall affect . . . the rights of any person in or over any land acquired before the commencement of this Ordinance."[80] Yet the law also gave the trustees power to make regulations concerning hunting and a variety of land use practices, which regulations would almost certainly interfere with existing rights. Obviously, the ambiguity of the law had the effect of fanning the flames of discontent as administrative officers would tell park residents that their rights were fully protected, while park authorities would try to enforce regulations to restrict their activities. The government's solution was to pass an amended law in 1954 that removed the contradiction by revoking the saving clause and replacing it with one that expressly denied any right of occupants to cultivate, and that gave the governor extraordinary powers to prohibit any other activities deemed undesirable. This did not diminish the conflicts.

Something clearly had to give. In the midst of the political unrest in East Africa and anti-imperialist criticisms from Europe and North America, Serengeti gained international notoriety. Subsequently in April 1956 the government published a report on the problems, recommending that Serengeti be reconstituted so that wildlife conservation and human interests be spatially segregated.[81] Once again, conservationists from England and North America intervened. The new recommendations had startled conservationists because the suggested area for wildlife was much too limited to contain the mass migrations of the plains ungulates. Their argument against the new boundaries was founded on a rejection of most African land claims in Serengeti. In a petition to the secretary of state for the colonies, a coalition of North American conservation organizations argued: "The inherent rights claimed by those who speak for the Masai are apparently neither defined in law, described in treaty, nor etched in history. An inherent right of these fine people, as well as others in Africa, by pure logic would extend to all land in East Africa, as well as the Serengeti Plains. We argue that the basic inherent right of the African is to have his natural heritages protected and defended even from his own errors."[82] In other words, if indigenous land claims were recognized in Serengeti, they would have to be recognized everywhere, including, as the authors seemed to imply, Kenya's White Highlands, where Europeans and Africans were engaged in a guerrilla war over land. Continued colonial rule, including the clearing of land of people for nature preservation, was ultimately justified as being in the best interests of the African population.

The government eventually was forced to appoint a "Committee of Enquiry" to review the issues, examine the various proposals for reconstituting the park, and make recommendations.[83] The Fauna Preservation Society (FPS; formerly the SPFE) sent their own biologist-consultant, whose report provided the basis for much of the committee's findings (Neumann 1995b). At the heart of the committee's recommendations was an endorsement of the principle that *human rights should be excluded in any national park*. The committee recommended that the national park should be reconstituted in the western Serengeti, and that the Ngorongoro Crater sector be excised from the park and managed as a special conservation unit where Maasai pastoralists would be allowed to stay. In a final sessional paper, the boundaries were refined based on consultations with the Maasai, and it was suggested that the necessary amending legislation be enacted.[84] In 1959 the National Parks Ordinance (Amended) was passed. Summing up the legislation, the chairman of the board of trustees wrote: "Under this ordinance the Tanganyika National Parks become for the first time areas where all human rights must be excluded thus eliminating the biggest problem of the Trustees and the Parks in the past."[85] A critical refinement of the national park concept had been made: there would be no place for people in Tanzania's parks. The policy had been established just as the sun was finally setting on the British Empire. The nationalists in Tanganyika had already reached agreement with the imperial government for a peaceful transfer of power. Recognizing the passing of an age, the SPFE had switched to a humbler appellation, the Fauna Preservation Society.

They remained keen on keeping the conservation movement going in the newly independent country, but other international conservation organizations had arisen after World War II and their involvement was rapidly overshadowing that of the FPS. The number one concern that dominated their meetings far away in Europe was the question of conservation in postindependence Africa. Could the new governments be trusted to keep up the conservation programs, and were they even capable of protecting the wildlife and wild habitats of the continent?

"Africanizing" the Parks

During the colonial period, little attempt was made to involve Africans in the management and conservation of the country's natural resources. Or rather, to put it more accurately, much effort went into *preventing* their involvement. Forest guards and game rangers were hired locally and a system of Native Forest Reserves was established in the 1930s, but these certainly didn't prepare Tanzanians to operate a state ministry of natural resources. The Africans whom colonial professionals had tried so hard to exclude were suddenly going to be in charge of the system of parks and reserves. John Owen, director of Tanganyika National Parks during the 1960s, acknowledged that park advocates "were tempted to overlook the importance of encouraging humans, particularly the local residents, to visit and enjoy the parks" (Owen 1962). A new approach was clearly called for, one linked to an invitation to the new African governments to join a global community of modern nation-states. Presenting wildlife conservation as a choice for African governments between civilization and savagery, Julian Huxley wrote: "In the modern world, as Africa is beginning to realize, a country without a National Park can hardly be regarded as civilized. And for an African territory to abolish National Parks already set up or to destroy its existing wild life resource would shock the world and incur the reproach of barbarism and ignorance" (Huxley 1961, 94).

The International Union for the Conservation of Nature and Natural Resources (IUCN) was founded in Switzerland after World War II to encourage and coordinate environmental conservation on a global scale. A major challenge was to assist the decolonized nations to plan and manage their own national parks and conservation programs. At the IUCN General Assembly and Technical Meetings in 1960, the delegates conceived of the "African Special Project," which would focus world attention, and the attention of the new leaders, on conservation in Africa. "The special purpose of the project," read the project committee's original report, was "to inform and influence public opinion through its leaders and responsible persons in Governments" that conservation was in the best interests of their nations (Watterson 1963, 9). The IUCN delegates felt that, worldwide, "the accelerated rate of destruction of wild fauna, flora and habitat in Africa . . . was the most urgent conservation problem of the present time" (10). After sending a representative to the governments of a number of African countries to elicit support, the IUCN organized a Symposium on the Conservation of Nature and Natural Resources in Modern African States. Newly independent Tanganyika, long the symbol of European conservationists' dreams and nightmares, would host the conference.

The conference was held in Arusha in September 1961, attended by representatives from twenty-one African countries and five international organizations. The fact that the conference was held in Tanganyika had implications reaching far beyond its symbolic importance. Governor Richard Turnbull's address at the opening session outlined the twin pillars of the government's wildlife conservation policy: "to develop a tourism industry and . . . to persuade public opinion [among rural Africans]."[86] Julius Nyerere, then Tanganyika's first prime minister (and later, first president), also addressed the meeting, encouraging international conservationists in a dramatic speech, now known as the "Arusha Manifesto." The section often quoted by conservationists reads: "In accepting the trusteeship of our wildlife we solemnly declare that we will do everything in our power to make sure that our children's grandchildren will be able to enjoy this rich and precious inheritance." This passage is followed immediately by an invitation to outside intervention: "The conservation of wildlife and wild places calls for specialist knowledge, trained manpower and money and we look to other nations to cooperate in this important task." Conservationists were delighted by the manifesto, and it continues to be cited in their documents and publications as a positive example of African government interest and cooperation in protecting wildlife. Usually going unmentioned, however, is the fact that it was written for Nyerere's speech by members of Western conservation organizations (Bonner 1993, 65). Nyerere gave his blessing to groups such as WWF and the IUCN to continue their involvement in establishing, planning, and managing the country's protected areas, largely to generate foreign exchange by developing its tourism industry. Education programs would convince the peasants and pastoralists who supported Nyerere's nationalist Tanzanian African National Union (TANU) party that parks and nature tourism were to their "own ultimate advantage."[87]

Tanganyika's minister of lands and survey once remarked that "it must, however, be said that the almost mystical and romantic regard for wild animals which some people have, has often puzzled the peoples of Africa" (Tewa 1963), punctuating Lusigi's (1984) observation that of all the inherited colonial institutions, wildlife conservation was the least understood within African culture. There was much discussion at the Arusha conference and elsewhere of educating the masses about wildlife conservation to rectify what was neglected under colonialism. Conservationists were encouraged to "work among the masses with missionary zeal" (Badshah and Bhadran 1962, 23) and "to awaken African public opinion to the economic and cultural value of their unique heritage of wildlife."[88] Beyond the obvious, these quotes inspire striking parallels with the efforts of early Christian missionaries, particularly their ideas about Africans as "natural Christians" (Curtin 1964, 225). Likewise it appears that Africans were regarded as "natural conservationists," possessing some latent capacity for appreciating European concepts of nature preservation that merely needed to be awakened. The implicit assumption was that Western values associated with "natural scenery" and "wildlife spectacle" were universal values.

Conservationists attempted this "awakening" of African values in contradictory ways—that is, through the use of symbols and media that were layered with meanings and values specific to European culture. One particularly imaginative, if not bizarre, project involved the production of a documentary film, *Safari to America,* which depicted a Maasai schoolboy's visit to Yellowstone National Park. Park advocates "hoped that the people of Tanganyika would find a record of this visit both interesting and instructive, and that it would bring out to them the cultural and economic importance attached to a National Park by even the most highly-developed nations in the world."[89] As another example, a poster captioned "Our National Parks are the envy of the world—be proud of them" was developed for mass circulation in 1961.[90] This single project embodies layer upon layer of unintended irony, beginning with the fact that a French artist was commissioned to produce the poster. It is equally ironic to use the theme of "national pride" to promote conservation among a population that had been exposed to the concept of African nationalism for less than a generation.

It seems especially dubious if we recall that TANU organizers adopted an anticonservationist platform in the rural areas to drum up support for their nationalist cause less than a decade before. In retrospect it appears that conservationists' efforts at "getting the local people into the parks so as to dispel any feeling they may have that the latter are run only for white men" (Owen 1963) were ill-conceived.

In any case, mass education was only a part of international conservationists' plans to deal with decolonization, and not the most important. Even they were skeptical of the power of their own message, recognizing that "those in daily contact with animals, whether they are persons whose crops are raided by elephant or hunters hungry for meat, are stony soil for general propaganda." Hence, it was "to the leaders at all levels of society and to the coming generation of leaders, that the main approach must be made" (Curry-Lindahl 1963). The African Wildlife Leadership Foundation (AWLF; now African Wildlife Foundation) was established during the era of decolonization specifically to address the training needs of African conservation professionals. At first, recruits were sent to the United States for training, but in 1963 AWLF, along with the World Wildlife Fund, helped to establish and fund the College of African Wildlife Management (CAWM) at Mweka, Tanzania, just an hour from Arusha. The guiding concept for CAWM was and continues to be to produce technically trained field officers for such positions as park warden for all of anglophone Africa.

Still, the problem of localizing park management was not easily solved. In 1967 the director of National Parks, a British expatriate, complained of the difficulty in finding "suitable men for such field appointments as Park Warden."[91] Three years later, the Tanzania National Parks Board of Trustees held an emergency meeting to address the problem. A subcommittee was formed that devised a plan to speed up localization,[92] whereby Mweka graduates would form "the backbone of our local service in the Parks."[93] Finally, on 1 January 1971 the first Tanzanian director of national parks was appointed.

The transition to African management had been made through an almost total reliance on CAWM at Mweka[94] and thus a reliance on international conservation organizations who continue to fund it, albeit with less direct involvement. The United States National Park Service has also played an important, though less visible role, sending planning consultants and hosting visits by Tanzanian park officials to American national parks. As a result, an elite class of bureaucrats, trained in Western ideologies and practices of natural resource conservation, has

emerged, much as it has in other state institutions in Tanzania (see Fortmann 1980; Hyden 1980). The continuity and connection between
colonial natural resource professionals and those of the independent
government are, in sum, quite direct. This colonial legacy of state-
directed conservation continues to influence contemporary relations
between park officials and local communities.

Economic Development, Preservation, and Dislocation

However unprepared Tanzania was to manage its
national park system, the new government was supportive of nature
protection. Whether Nyerere's support of nature preservationists'
agenda reflected the new government's desire to earn the respect of the
international community, a sincere scientific interest in conservation, or
any number of other moral, political, or scientific objectives, is difficult
to say. It is clear, however, that the government expected the parks not
only to pay for themselves but to provide the state with a major source
of foreign exchange. TANAPA's director John Owen recognized this,
and under his guidance the national parks agency would "encourage by
every means the growth of a tourist industry" (Owen 1962). Previous
five-year plans for national development have projected that tourism,
largely based on the attraction of the parks, would provide the country's
second largest source of foreign exchange. Because of their expected
role in fueling economic growth, the administration of national parks
was transferred in 1968 from the Ministry of Agriculture, Forests and
Wildlife to the Ministry of Information and Tourism.[95] For the government, park tours became an exportable commodity with great
potential to fuel state accumulation. To this point, an official wrote
that the purpose of the parks "is the earning of foreign exchange, in the
same way that one looks upon the exports of coffee, sisal, cotton, tea or
diamonds."[96]

Tourism is only now beginning to live up to its promise as a revenue
generator (Neumann 1995a). Through the mid-1970s, Tanzania experienced a fairly steady growth in the number of visits to its national
parks, averaging an 11.1 percent increase per annum from 1969 to
1976 (Curry 1982, 14). However, the government-owned Tanzania
Tourist Corporation (TTC) was invested heavily in tourist facilities,

building four new hotels in northern Tanzania that operated at a loss into the 1970s. From the late 1970s, revenue earned from national parks stagnated and then declined, until 1986 when it began a steady resurgence (United Republic of Tanzania 1988). Many factors had kept tourism potential down through the years, but one of the most important was that neighboring Kenya's industry is highly developed and is supported by a relatively advanced infrastructure necessary to support mass tourism. During the period of the economic union of Uganda, Kenya, and Tanzania as the East African Community, Tanzania received little of the overall tourist revenue for the community, even though parks such as Serengeti and Kilimanjaro were major destination sites. Most of the safaris were arranged, outfitted, and paid for in Nairobi before they entered Tanzania, a situation that has existed for decades.[97]

The establishment of a tourist industry has been a major motivating force behind Tanzania's expansion of its system of national parks. In 1969 the TANAPA director wrote, "[S]ince independence Tanzania has quadrupled her expenditure on National Parks. In percentage terms this represents a higher proportion of the national budget than that allocated by the United States for its National Parks[,] . . . mostly due to the fact that the tourist industry is expected to be the prime earner of foreign currency for the country" (Owen 1970). The expansion of wildlife conservation areas was thus explicitly linked with national economic development. Nyerere himself was quite unambiguous in his position on the role of wildlife in bringing in foreign exchange: "I personally am not very interested in animals. I do not want to spend my holidays watching crocodiles. Nevertheless, I am entirely in favor of their survival. I believe that after diamonds and sisal, wild animals will provide Tanganyika with its greatest source of income. Thousands of Americans and Europeans have the strange urge to see these animals" (quoted in Nash 1982, 342). Politically, this plan carried potential hazards, since many Africans saw tourism as exhibiting the worst aspects of the old colonial order of European dominance and African subservience (Shivji 1973).

Recent changes in Tanzania's political economy have sharpened the tensions between rural populations, the tourism industry, and protected areas (Neumann 1995a). The seeds of change can be traced to 1985, when Nyerere resigned from the presidency and his successor, Ali Hassan Mwinyi, agreed to adopt the International Monetary Fund's (IMF) structural adjustment program. At that time, the tourism industry had been controlled by a government parastatal, the TTC, which owned and operated fifteen hotels and lodges in the national parks and elsewhere. Under the IMF-induced privatization program, the TTC was dissolved in 1993 and reconstituted as the Tanzania Tourist Board (TTB), a promotional office for private investors. The dissolution of the TTC combined with a new investment code has resulted in a massive infusion of foreign capital in tourism. In 1996, for example, Kenya-based Sopa Lodges completed a $7 million luxury hotel in Tarangire National Park, its third new resort in Tanzania in three years (EIU 1996). Tanzania's new five-year development plan for tourism, unveiled in 1996, projects a doubling of the number of tourist arrivals in the coming decade. As a new phase of accumulation emerges in the industry, communities surrounding protected areas feel increasing threats to their land claims and loss of economic opportunity, since the bulk of profits are captured by foreign-owned companies (Neumann 1995a).

The recent growth in tourism in Tanzania is properly seen as a spike in a long history of expanding nature protection and the development of associated infrastructure and international intervention. Throughout the three decades of its independence, the government, with help and persuasion from various international organizations and foreign government agencies such as the U.S. National Park Service (Tanzania National Parks 1973), continued to gazette national parks. Sometimes, as in the case of Arusha National Park, international organizations would actually provide the money to buy the land to be protected. Often, however, the land was already under some form of state protection. While it is frequently noted (perhaps in part to downplay the ties to colonial policy) that all but one national park, Serengeti, were created after independence, this observation is somewhat misleading. Commonly the boundaries of national parks are based upon areas placed under protection by the colonial powers as forest or game reserves (see table 3).

Though the details were poorly documented (Pullman 1983), mass relocation was often necessary for the establishment of such protected areas, from the colonial period and continuing into the present (see table 4). Sometimes relocations from protected areas overlapped with other government agendas. Former President Julius Nyerere, for example, supported the evacuation of Serengeti National Park partly because it coincided with the objectives of the government's villagization strategy (Tanganyika National Parks 1964). Earlier colonial-era relocations occurred as part of larger government efforts to control rural populations. The most notable example is the Selous Game Reserve, the

Table 3 *Examples of Postindependence National Parks Created from Colonial Reserves*

Colonial Designation	Date Established	Current Designation
Gombe Game Reserve	1943	Gombe National Park
Katavi Plain Game Reserve	1928	Katavi National Park
Kilimanjaro Game Reserve	1896	Kilimanjaro National Park
Lake Manyara Game Reserve[a]	1957	Lake Manyara National Park
Mount Meru Forest Reserve	1908	Arusha National Park
Rubundo Island Forest Reserve	189?	Rubundo Island National Park
Saba River Game Reserve	1910	Ruaha National Park
Serengeti Game Reserve	1908	Serengeti National Park
Tarangire Game Reserve[a]	1957	Tarangire National Park

[a] All or part of these areas were designated earlier as Game Controlled Areas

largest protected area on the continent. The German colonial administration initially established a game reserve in the area in 1905. In the same year, the Maji Maji uprising—the most extensive African rebellion against colonial rule in Tanzania's history—originated here (Iliffe 1969), sparked in part by the severe restrictions imposed by German game laws (Koponen 1995). The reserve was later expanded by the British administration in the 1940s as part of a scheme to depopulate large areas of southeastern Tanzania in an attempt to isolate tsetse fly, the vector for sleeping sickness (Matzke 1977). As the reserve was expanded, successive governments evicted an estimated total of forty thousand residents from within its boundaries (Kjekshus 1977; Yeager and Miller 1986).

The Maasai have probably been the most severely affected by the establishment of protected areas in East Africa. In the Serengeti region alone their grazing area was reduced from forty thousand square kilometers to sixty thousand square kilometers (Kjekshus 1977). In a 1990 paper, Henry Fosbrooke—a former conservator of Ngorongoro Conservation Area, whose familiarity with pastoralists began when he was appointed as district officer in Tanganyika Maasailand in 1934—recounts a relatively recent eviction in Ngorongoro Conservation Area: "Early one morning in March 1974 three Land Rovers entered the

Table 4 *National Parks and Game Reserves Established by*
        Population Relocation

Protected Area	Number of People Removed	Source
Burigi G.R.	?	Rodgers et al. 1977
Gombe N.P.	500	Kjekshus 1977
Katawi G.R.	?	Kjekshus 1977
Mibulu G.R.	10,000	Kjekshus 1977
Mkomazi-Umba G.R.	5,000	Fosbrooke 1990
Saba River G.R.	?	Kjekshus 1977
Selous G.R.	40,000	Yeager and Miller 1986
Serengeti N.P.	1,200	Arhem 1984
Tarangire N.P.	?	Mascarenhas 1983

Crater, one going to each boma. They carried personnel of the para-military Field Force Unit, F.F.U., termed in Swahili, *Fanya Fujo Utaona* (If you make trouble you will see!). Without explanation and without notice they ordered the immediate eviction of the inhabitants and their cattle. Their possessions were carried out by transport of the Conservation Authority and dumped on the roadside at Lairobi. No explanation was given and no arrangements made for the re-settlement of the evacuees" (Fosbrooke 1990).

As an illustration of how this practice continues, in 1988 more than five thousand pastoralists were forced out of the combined Umba and Mkomazi Game Reserves after refusing to obey a government eviction order.[98] The original 1951 order, which created the reserve, legally guaranteed the residents of the area the right to continue living and grazing livestock there. By the 1970s, wildlife officials were claiming that the Mkomazi had become overrun by pastoralists and their cattle, although this is strongly disputed (Fosbrooke 1990). In 1976 the reserve manager informed the pastoralists that they must move out. Resident Maasai pastoralists' efforts to get the Mkomazi eviction decision reversed were answered with threats from armed Game Department agents (Mustafa 1993). Finally in 1987 a directive from the Ministry of Natural Resources and Tourism canceled all previous permits for grazing and residency (ibid.), and in 1988 all residents were evicted from Umba-Mkomazi, some by force after refusing to leave voluntarily.

The strategy to promote wildlife preservation through coercion has not helped the conservationists' battles to "educate" the masses. Local communities have historically resisted these dislocations. Government park reports from the 1960s inevitably contain references to confrontations between park officials and local residents (United Republic of Tanzania 1964; 1965; 1966; 1967). Illegal settlement, grazing trespass, and cultivation encroachment were (and remain) a perennial problem for park administrators. Management strategies ranged from arrests (Republic of Tanzania 1966) to bulldozing the park periphery to make the boundary more visible (Tanzania National Parks 1970). Resistance to the establishment of Biharmulo Game Reserve delayed its establishment for four years, after twice forcing the readjustment of proposed boundaries (Rodgers et al. 1977). Incidents such as these have produced a lingering cloud of animosity and discontent between park officials and nearby communities that has plagued nearly every protected area in Tanzania.

The process of park establishment acted out at Serengeti was thus essentially repeated, albeit with variations, throughout Tanzania after independence. Once the human rights issue was legally clarified, customary claims were either ignored or curtailed, and the relocation of human populations to create parks elsewhere could take place, legitimized by written law. Not all cases of park establishment, for a variety of historical reasons, were so directly disruptive to existing land use practices. For instance, the first postindependence national park, Arusha, had its own unique history of human use and occupation that initially dampened direct confrontation over its establishment.

The Creation of Arusha National Park

Like most of Tanzania's protected areas, the origin of Arusha National Park is linked directly to the country's colonial past. With the exception of a small portion of former public land, the park was established on land that was either colonial forest and game reserve or alienated for European estates. In fact, two parcels of alienated land later included in the park—known as Momella Farm—were technically a protected area, as the owners (the Trappe family) had the government legally declare their land a hippo reserve in 1931. Before the arrival of the Europeans, the estates and reserves were utilized or occupied by the Meru and Maasai peoples. Today the park lies entirely within the area historically claimed by the Meru and used as a grazing commons (some of it shared with the Maasai) and as a source of numerous products from the forest until the colonial governments outlawed most uses in the name of conservation.

Mount Meru had long been considered an ideal candidate for a national park as a result of both its spectacular 4,565-meter volcanic peak and its subsidiary Ngurdoto Crater, which provided habitat for herds of rhino, elephant, and buffalo. Commenting on Major Hingston's 1930 report on wildlife conservation in Tanganyika, the game warden asked that Mount Meru be included in the major's list of potential national parks,[99] but it was not until after the conflict at Serengeti had been dealt with that any action was taken. The idea for creating a national park at Ngurdoto Crater received a boost from the publicity surrounding the 1958 visit of Britain's Princess Margaret (Vesey-FitzGerald 1967). Gazetted in 1960, the minuscule Ngurdoto Crater National Park (697 hectares) became the second national park in the country and technically the first postindependence park. As such it carried some symbolic weight among international conservationists who took time out from the Arusha Conference to attend an official opening in September 1961.[100]

Because Ngurdoto Crater was part of the forest and game reserve, it was assumed the Meru had given up most of their rights long ago.[101] In fact Meru tribal authorities were consulted before the creation of the tiny park,[102] which included only the crater up to its rim. Since it was first gazetted, however, the boundaries of the national park have been extended by four separate legislative acts (see map 6). One of World Wildlife Fund's first projects in Africa was the purchase of Momella Farm for inclusion in the park, made possible by a significant contribution by a German woman interested in protecting the area.[103] Together with a piece of public land that included portions of Momella Lakes, the farm became part of the park on 19 October 1962.[104] The Meru Native Authority land subcommittee was consulted, and it agreed to the new boundary in exchange for the larger portion of the public land plot.[105] Following this the Forest Department turned over a portion of the forest reserve lying between Ngurdoto Crater and Momella Lakes, and this was gazetted on 18 September 1964. The park's area was increased from thirteen square kilometers to fifty-two square kilometers.[106] Then in 1967, a second national park, Mount Meru Crater, was gazetted, consisting of 4,905 hectares on the mountain's uppermost slopes, and the two parks were combined five months later to form the 9,675-hectare Arusha National Park. The expansions continued in 1969 when

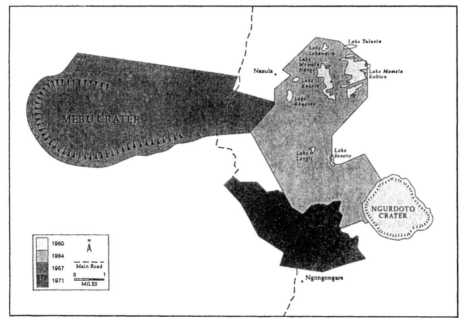

Map 6. Arusha National Park expansions. (R. P. Neumann, compiled from legislative descriptions.)

the national park benefited from the nationalization of several European estates, picking up an additional 1,895 hectares. Once again, the funds to purchase land from the park came through an international conservation organization. As noted in a TANAPA report, "Our boundaries were enlarged by the taking over of the Ngongongare farm by the Government. This was given to the Arusha National Park to administer, pending its gazetteering, together with the Trappe farm. Funds for the purchase and development of both Ngongongare and the Trappe farm were provided by the African Wildlife Leadership Foundation of Washington, D.C.[,] to whom we extend our grateful thanks."[107]

The historical development of the park, then, consisted of incremental expansions, with wealthy donors from Europe and North America providing funds to international conservation organizations to purchase the estates of departing European settlers. Thus, as the park stands today, it encompasses a small but ecologically diverse area ranging from alkaline lakes to closed-canopy montane forests to the subalpine mountain summit. Additionally—and this is most critical for explaining the conflicts between the park and surrounding communities—a land use

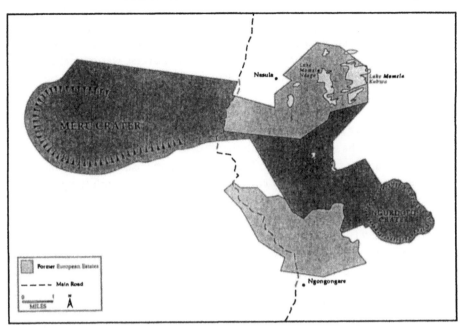

Map 7. Location of former European estates in Arusha National Park.
(R. P. Neumann, compiled from legislative descriptions.)

pattern was established whereby the park areas now bordered by Meru
settlements are precisely those areas that were once alienated to European settlers (see map 7).

ELIMINATION OF REMAINING RIGHTS

Though most of Arusha National Park's lands had been
taken out of local control and customary practices were outlawed
decades before the park was established, a few rights still existed within
the new boundaries. These were soon eliminated. The right to collect
wood in the Meru Forest Reserve for household use, for example, was
rescinded soon after the land was transferred to TANAPA authority.
The 1964 and 1967 acts, however, contained an exclusionary clause
that protected a right of great local importance. A three-meter-wide
right-of-way bearing southwest by northeast through the section of former forest reserve was excluded from the national park to allow the
Meru to continue moving their cattle and crops across the mountain.
By all accounts it was an ancient and major route for trade and the

movement of cattle, not only for the Meru but the Maasai and other peoples as well. The path that bisected the forest reserve was part of a larger route directly connecting Maasailand in Kenya with the trading center of Arusha town and points beyond. The access was critical to Meru peasants living nearby as most had family and many had farms on both sides of the park.

The importance of the right-of-way was recognized at the early stages of planning for the park's establishment. In 1956 the district commissioner of Arusha wrote of the path that "there is a great need for access for the Wameru from Kilinga to Leguruki[,] and the team feels that this aspect ought to be very fully considered in conjunction with the proposals for the proposed National Park."[108] The park administration was less than enthusiastic about the idea, and the director advised the board of trustees, "The Provincial Administration wish to legalise this right of way and the Game and Forest Departments have no objections. Should the area be made a National Park, however, the Trustees would have the very strongest objections to the passage of humans and stock in any part of the Park, and particularly here, where the route lies at right angles to the movement of game and cannot therefore be fenced."[109] He noted further that "[t]he track is bordered on both sides by grassland which is now illegally grazed by Wameru stock[,] and the declaration of the area as a National Park is unlikely to alter the Meru attitude to what has become a 'right' in this respect." Nevertheless, TANAPA was forced to agree to the right-of-way or give up the idea of a contiguous park boundary enclosing the upper slopes of Mount Meru.

The local park authorities did not let the matter rest there. Apparently sometime in 1973 the park began to prohibit villagers from using the path.[110] It is not clear that anyone, either the park administrators or the villagers, realized the existence of the clause at the time, though villagers today claim that the path was not closed through any legal procedure, but only by means of intimidation by park guards. For their part, park administrators seem to have been caught totally off guard by the existence of the right-of-way. "In my investigations," the Arusha National Park warden wrote, "I have discovered unexpectedly that legally people had a path allowing them to pass in the park coming from Seneto going to Ngongongare or vice versa. I am at a loss to come upon unexpectedly that citizens truly still have the right of passing inside the park legally. Until now we have not given any person at all permission to pass this way."[111] The director confirms his subordinate's unsettling discovery but advises him that it might be "best to 'let sleeping dogs lie.'"[112] The park effectively splits Meru land into two separate zones, but the path allowed people to traverse the park in about two hours. With the closure of the path, a two-hour journey was turned into one that could barely be completed in a twelve-hour day.

The right to keep bees in the forest, a practice that was controlled by the Forest Department through a permit system before the park took control, was also gradually eliminated. After the gazetting of the park, officials restricted the beekeepers' movement in the park and required that they be accompanied by a ranger whenever they entered to check their hives. "Few licensed bee-hive keepers now come into the Park to inspect their hives," wrote a park warden in 1969, "as they have to be accompanied by a Ranger. It would appear that previously they had not confined their activities to bee-keeping" (Tanzania National Parks 1969). The warden did not elaborate on his innuendo, but he seemed to imply that honey-gathering was a ruse for hunting park animals, and with a ranger along it was no longer worth the beekeepers' time to enter the forest. The (former) beekeepers expressed a different interpretation of the situation, complaining that it was often difficult to make arrangements to get an escort because it was a low priority for the guards.[113] In 1976, a generations-old practice on Mount Meru was ended when they were told by the warden to remove their hives from the park; according to the beekeepers, the warden believed that the guards were spending too much time away from their other duties.

In addition to the material benefits that the Meru enjoyed, Mount Meru also holds ritual importance for two of the larger clans. Below the crater of Mount Meru lies the grave of Lamerei, the Meru ancestor of the Mbise clan who was the first to arrive on the mountain many generations ago. It is the site (called Njeku) of what had been, until recently, an annual ritual for the Mbise and Nnko clans to end the dry season and call forth a new season of rains (see figure 8). But as an elder involved with conducting the ritual explains, the importance of the practice is changing, partly because of changing religious practices and partly because of interference from the park. "We used to go up once a year," the elder recalled, "usually in January or February. Two years we have not been there and this will be the third year because we will not go. Since the last time, all of those wardens [the speaker names them here] have not allowed us to go there. They say that in the park no person at all is allowed to walk without a park guard. But our custom does not allow us to take anyone along on our prayers. At present there are very few of us remaining: less than ten and we need more than thirty.

Figure 8. A clan elder stands before a giant cedar at the customary camp at Njeku, the ancestral site in Meru Crater. (R. P. Neumann.)

The major problem is that we can not go there without these children and all the children are baptized. Even me, I am very close to being baptized."[114]

The park administration has labeled Njeku a "cultural site" and has plans to develop it as an attraction for foreign tourists. In sum, Meru communities have experienced a progressive erosion of whatever rights and privileges remained from the colonial era, while simultaneously watching the area under park control expand exponentially.

Among conservationists, natural-resource conflict in Africa is frequently seen as being driven by population growth, poverty, and an ignorance of conservation principles. While all of these factors may play important roles, the repression of certain events in the collective conservationist memory—the eradication of customary rights, forced relocations, and broken promises—inhibits a full understanding of the conflicts. The historical picture of conservation that emerges is thus critically flawed. The construction of this history requires the suppression of the voices of the peasants and pastoralists who have borne the costs of state-directed conservation. Once their productive activities were removed from the new landscapes of consumption, they disappeared also from the pages of history. In the most recent tourist guide to Arusha National Park (Snelson 1987), the Meru are not even mentioned in the section on the area's history, which, instead, relates a story of the first European to "discover" the mountain. If the fact that customary rights preexisted the establishment of protected areas is recognized at all, these rights are "of minor importance,"[115] making it easy to create parks with "minimum disturbance to existing rights."[116] Displaced peoples become poachers and bandits, a Malthusian nightmare of humanity waiting to invade the protected lands.

A quick examination of a speech given at the Serengeti and Ngorongoro Diamond Jubilee celebration can illustrate the subtle ways that the history of wildlife conservation is constructed and subsequently influences our understanding of natural resource conflict. The following statement introduces the speaker's brief history of the division of Serengeti into a national park and separate conservation area: "As far back as 1921, up to 1959, for thirty eight years, Ngorongoro and Serengeti were one thing known as 'Serengeti National Park.'" The statement is erroneous and misleading. Serengeti National Park did not exist until the passage of the 1940 Game Ordinance; much of Ngorongoro had previously been a German cattle ranch and was not designated a game reserve until 1928. The implication of this statement and the historical account that follows it is that the park was solidly in place for years until the pressures from local communities—"poachers" and "encroachers"—forced its split. There is no reference in the speech to the numerous guarantees of rights that were made to residents, nor to the subsequent elimination of those rights and removal by force of those resisting their loss.

When the loss of access rights and the human costs of nature protection *are* acknowledged, they are expressed within a let-bygones-be-bygones sentiment. "As the Arabs say, *el fat mat*—the past is dead," wrote a former TANAPA director. "Let us concentrate on the future[,] recognizing that what has gone before casts its shadow along the path ahead—that we have not only to awaken public opinion, but to change it." (Owen 1963, 262). A strikingly similar attitude was expressed to me in a conversation with a leading Tanzanian wildlife official in 1990. "What's done is done," he said. "Sure people lost land, but it is time to move on." The problem with this position is that it is seen by the conservationists as the starting point for negotiations with displaced communities. It remains to be seen whether the people who have lost their land and resource access will so readily agree to this position.

The direct confrontations over the establishment of Serengeti National Park were not replicated on Mount Meru, since most rights had been outlawed decades before and whatever rights remained were whittled away piecemeal rather than being eliminated overnight by the Arusha National Park Act. In fact, some customary practices, such as beekeeping and the movement of cattle and people through the protected area, were permitted and even guaranteed by law, thus initially avoiding difficulties over the park's creation. Indeed, even explicitly illegal acts by the Meru were tolerated in the early days of the park, presumably to reduce friction with local communities: "A constant educational-type battle was waged against illegal grazing of cattle, illegal bee keeping and trespass by the local Wameru. No cases were prosecuted as a matter of policy."[117] Nevertheless, as we shall see, many of the historical patterns of struggle between the Meru and the state over access to protected lands and resources, and over their meaning, persist.

5

Patterns of Predation
at Arusha National Park

The current pattern of land use on the upper slopes of Mount Meru is the result of historical processes that have produced settlement, cultivation, and grazing adjacent to portions of an extremely irregular park boundary. In other locations, the intensive land use borders the Meru Forest Reserve, which lies between the villages and the national park. The lack of attention to and planning for adequate habitat to contain the park's wildlife populations, and the concentration of landholdings dating to the colonial period, combined with population growth and the expansion of the park's boundaries to adjoin existing agricultural areas, have all contributed to the current situation. As a result, Arusha, perhaps more than any other park in Tanzania's system, is plagued by a continuous clash of wildlife and human interests: park animals are illegally hunted, livestock trespass into the park, wildlife destroy crops and threaten human life, and diseases are passed between wildlife and livestock.

This chapter details how the conflict is manifested for parties on both sides of the boundary, what it means for the park administration's goals of protecting flora and fauna, and how it affects agricultural production and daily life in the villages. It will reveal how the specific pattern of conflicts manifested day to day is a reflection of the complex interaction of ecology, the seasonality of peasant production practices, and local economy.

Arusha National Park Ecology

The ecological history of the park is greatly influenced by the volcanism that created Mount Meru, and the soils to a large extent are a product of this activity. An estimated six thousand years ago (Lundgren and Lundgren 1972, 227), the eastern wall of the Mount Meru caldera collapsed, releasing a gigantic landslide that spread twenty-four kilometers to the east and south and produced the hummocky terrain surrounding Momella Lakes known as lahar topography. The depressions are commonly filled by lakes or swamps, many of which are alkaline. Soils in this area, which includes the Nasula community, are shallow, rocky, and relatively infertile, with occasional pockets of more developed soils. Areas on the mountain that were not affected by the catastrophic collapse of the caldera are characterized by deep, unconsolidated ash deposits.

The extreme ranges in elevation and the effect on spatial rainfall patterns resulting from the interaction of climate and relief make for substantial ecological variability within the park. Elevations range from 4,565 meters on the rim of Mount Meru Crater to 1,448 meters at Momella Lakes, and rainfall varies from an annual average of 127 centimeters near Ngurdoto Crater to 89 centimeters around Momella Lakes, peaking at 229 centimeters at the higher elevations. Rainfall is monsoonal, with most of it falling during the movement of the Intertropical Convergence Zone (ITCZ) across the area: northward with the "long rains" in late March and April, then returning southward in November and December. The atmospheric uplift caused by the mountain directly affects precipitation, which increases with elevation up to 2,438 meters and then declines. The mountain also deflects the prevailing rain-bearing winds, thus determining rainfall distribution. The southeasterlies bear the long rains, and the south and southeastern sides of Mount Meru therefore receive the heaviest rainfall. The northern sides are significantly drier as the short rains produce less average rainfall or, not uncommonly, fail altogether. These extremes in climate conditions, combined with the influence of human activities and edaphic variability, have produced a vegetation mosaic in the park. Along the south side of the park, mostly in the area of former forest reserve between Ngurdoto Crater and Mount Meru, there is a thick cover of montane forest, while much of the rest of the park is vegetated by secondary shrubs or grassland (Vesey-FitzGerald 1973).

The grasslands within the park can be broadly typed as either edaphic or derived (Vesey-FitzGerald 1974). The edaphic grasslands are characterized by high water tables that keep the soil submerged for at least part of the year. The derived grasslands, the precise history of which is unknown (Lundgren and Lundgren 1972), are the result of past human activities. The oral histories produced by Meru informants indicate that cattle grazing and firing of vegetation, both by the Meru and Maasai, had probably been conducted for at least two hundred years before the Europeans arrived.[1] Pieces of clay pottery discovered in one grassland site within the park were found in soil deposits that were carbon-dated at A.D. 800–1200 (239), indicating that the high-elevation site has a history of occupation that predates the arrival of the Meru.

The areas of the park that were formerly European estates were managed in a way that had significant impact on the vegetation within the park. The European settlers used the most fertile areas to grow coffee and other crops while pasturing cattle on the rest of the land. Fires were encouraged as a means of improving grazing and, as a consequence, much of the vegetation consists of tussock grasses and secondary shrubs (Vesey-FitzGerald 1973). This, in essence, increased the habitat for buffalo, the principal grazing species in the park and, at least in the first decade of the park's existence, their population increased (1974). The grazing mosaic that developed as a result of fire and livestock was maintained during the sixties and into the seventies by wildlife species such as buffalo, eland, and elephant.

In recent years, however, grazing pressure from wildlife does not seem to be able to maintain the grassy glades in the park. A follow-up to Desmond Vesey-FitzGerald's study indicated that in the Momella-Nasula area, Momella Lakes, Ngongongare, and the basins of Lake Kusare and Lake Elkhekhotoito the ratio of herbs to woody vegetation had declined considerably (Mwijustya 1985). Mwijustya estimates, for example, that in the area of Momella Lakes the density of secondary shrubs had increased sixfold in the twelve years prior to his follow-up study. It is not entirely evident why this change is occurring, although it is likely a combination of the exclusion of fire and livestock and possibly a reduction of grazing pressure from wildlife species. The lack of empirical data on wildlife populations makes conclusions difficult. A population estimate for buffalo in the early seventies put their number at eighteen hundred (Vesey-FitzGerald 1973), but there has been no population study since then to indicate if the population is on the increase or the decline. Likewise for elephants, the other important grazing species, there are no population estimates, though qualitative assessments (Thorsell 1982) and anecdotal information suggest that their numbers, as well as those of eland, have been significantly reduced.[2]

In fact, drawing any conclusions about the ecological status of the park in general is problematic. No major ecological research has been conducted in the park since the early 1970s when Vesey-FitzGerald was stationed at the park as a government scientist. A study of the dynamics of the park's giraffe population conducted in 1979–80 predicted that the species' numbers were likely to decline (Pratt and Anderson 1982). For rhinoceros, also a major browsing species, only qualitative evidence for their decline exists. For example, seventeen animals were seen congregated in Ngurdoto Crater alone in 1961,[3] but there has not been a publicly documented sighting in the entire park since the mid-1980s. The extirpation of rhinoceros is the one conclusion concerning wildlife populations that can be made with any certainty despite the lack of empirical research. The management of the national park is conducted within this informational vacuum.

Park Management and Planning

The fact that much of the park is secondary vegetation presents some tricky management questions concerning what is being protected. The national park ideal is to preserve nature, a concept that implicitly assumes that the area to be protected is "natural," a difficult proposition on Mount Meru considering the long history of human influence on the land. An attempt by Arusha National Park's first and only scientific officer, Vesey-FitzGerald, to clarify the park's overall objectives alludes to the dilemmas and confusion surrounding the protection of nature there: "The objective is the restoration of a state of naturalness, but first of all it is necessary to establish what is a state of naturalness. It is not a pristine condition—no such has existed on this explosive landscape. It is not the man-degraded habitat that Tanzania National Parks acquired. It is a state of harmony between the existing topography and the existing plants and animals. It is an ever-changing state related to the erosion cycle, plant succession and the impact of animals on their environment. It is 'the balance of nature,' if you like, but the balance of nature swings."[4] This halting attempt to outline park objectives takes us to the heart of the contradictions presented when romanticized, ill-defined notions such as "harmony between topography, plants, and animals" and "balance of nature" collide with the reality of ecological change and the historical role of human society in shaping "nature."

Regardless of the difficulties in defining the term "natural," it is clear that the ecology of large portions of Arusha National Park was shaped by human activities. The gazetting of the park and the exclusion of humans, then, meant the removal of major ecological influences and subsequent changes in the park's ecology. Changes in vegetation mean changes in wildlife habitat, and some species benefit while others may eventually be excluded. The general strategy has been to "let nature take its course," except in the ironic case of vegetation clearing by the park staff in order to maintain opportunities for tourists to view wildlife.[5] It is ironic because the park staff is forced to clear brush once held in check with the help of livestock grazing and burning, two of the activities that the park administration has fought hard to exclude. Meru peasants living nearby are hired as casual labor by the park administration to cut and burn the glades that their parents and grandparents once maintained through their livestock-herding practices.

Beyond the sporadic effort to keep back the brush, the government has taken little direct action in ecological monitoring or management of the park. For most of its existence the park has been administered without a management plan, and day-to-day operations are guided by broad national park policies and the inclinations of individual wardens, who even into the early 1980s had not been trained at CAWM. Since TANAPA policy is to frequently rotate the assignments of its park wardens, park management strategy has been less than stable. Furthermore, management difficulties were compounded throughout the 1980s by a perennial fiscal crisis in the agency set against the backdrop of overall decline in the Tanzanian economy. The infrastructure of the park deteriorated and staffing was inadequate. The Arusha National Park warden's monthly report of October 1986 complained that most guard posts had only one ranger (the guard post at Nasula was not staffed at all for most of the year)[6] and therefore effective patrols could not be mounted.[7] The inadequate support for natural resource law enforcement meant that "the poachers are getting away," the warden wrote.[8] Without sufficient funds from the central government, the park was often left without a vehicle for lack of petrol or spare parts, and payrolls for what staff there was were not met.[9] A survey conducted at the park in the early 1980s found that "salary payments to park staff have been irregular with delays of up to three months experienced. Uniforms were seen on only two members of the staff. Morale and discipline problems have increased" (Thorsell 1982).

Since the mid-1980s, administrative conditions at the park have vastly improved. The budget for TANAPA has been increased considerably in recent years, and by 1990 Arusha had by far the highest expenditure per square kilometer in the eleven- (now twelve-) park system, nearly double that of the next highest. The size of the park staff has been increased and their level of training improved. During the period of my research, fifty-four rangers were on the payroll, and both the chief park warden and one of the assistant wardens had graduated from the program at CAWM.

Crimes against Nature

Not surprisingly, given the management situation, the late 1970s to the mid-1980s witnessed a dramatic decline in elephant numbers and the loss of rhinoceros to illegal trophy hunters. The Arusha National Park ranger and warden reports throughout the late 1970s describe finding shot rhino,[10] but usually after the horn was long gone, and rarely were there arrests associated with commercial poaching. In fact, the two times that the reports even mentioned suspects in the rhino cases, the blame was directed toward the forest guards and game scouts,[11] as in this account: "Also one rhino was lost near the rest house. It was believed that this rhino was shot by the game scout from Oldonyo Sambu[,] who was pretending that he had come to look after the villagers' maize crop."[12] Arusha National Park differs significantly from the more famous East African protected areas like Selous Game Reserve and Serengeti National Park in that it has not experienced the sort of large-scale organized poaching operations for rhino horn and ivory found in those areas. This is due primarily to the small numbers of these two species on Mount Meru relative to the populations of larger parks in savanna habitat. In any event, it is widely acknowledged (Thorsell 1982; IUCN 1987, 825) that most of the commercial poaching for horn and tusks was conducted largely by people from outside the area (i.e., non-Meru).

While the loss of these species is a significant threat to the park, the most persistent and widespread management problems relate more to the production and reproduction demands of surrounding Meru households. What little commercial poaching for horn and ivory there was at Arusha largely disappeared along with the elephants and rhinos. The natural-resource crimes that now plague the park administration are characteristic of the historical struggle over access to subsistence resources between Meru peasants and the state. The essence of the present situation is well represented in this 1979 log entry from the Rydon Farm Guard Post:

> 17/1: U. M.'s 35 goats were apprehended in the park. He was fined a total of 100 Tsh.
>
> 19/1: S. J. was found with four snares for trapping bushbuck. He was fined a total of 100 Tsh.
>
> February: calm
>
> 1/3: S. S.'s 14 cattle, 11 goats and 10 sheep were caught in the park at 5:45 pm. He was fined a total of 50 Tsh.
>
> 11/3: four dogs shot dead in Ngongongare area.
>
> 1/4: M. A.'s 31 goats were caught in the park. He was fined a total of 50 Tsh on 2/4.
>
> 18/4: M. N. was apprehended cutting grass in the park. He was fined a total of 50 Tsh. One dog shot dead.
>
> 25/4: N. N. was apprehended grazing his livestock in the park. He was fined a total of 50 Tsh on 27/4.[13]

Though the situation varies from month to month and year to year, this passage is fairly characteristic of the types of natural-resource crimes that occur most frequently. The crimes—fuelwood collection, trapping for meat, grazing trespass, and cutting grass for livestock fodder or roofing—are essentially attempts by Meru peasant farmers to obtain from the park some basic resources critical to household production and reproduction strategies.

At present the management's problems with illegal hunting are centered on the trapping and shooting of wildlife for meat, often for local sale rather than direct consumption. In 1980 the warden reported, "Poaching is increasing day after day. This includes the hunting of warthogs and buffalo for meat to be sold in the villages." A market for game meat has developed in the communities surrounding the park, principally giraffe and buffalo, but warthog, bush pig, and antelopes, such as bushbuck, also have market value.

The most prevalent method of obtaining meat is to set wire snares in the forested areas of the park. Essentially, the trapper studies the movement of animals along identifiable "game trails," sets the snare, and returns later, often under the cover of night, to recover whatever animal has happened to be trapped. Spears are also commonly used, though this requires a bit more skill and risk. Least desirable of all is the use of firearms—they are beyond most peasants' means, the bullets are expensive, and, in a small park like Arusha, almost sure to be heard. Some of the illegal trapping and hunting is motivated more by the need to defend crops against raiding wildlife than to obtain meat.[14] Since it is too destructive to set traps in one's plot (a buffalo thrashing around with a wire noose around its neck, dragging a heavy log, would wreak considerable havoc), farmers often set traps just inside the park boundary.

The social and ethnic composition of outlaw meat hunters is varied. Those arrested tend to be males from peasant backgrounds, below middle age, and with usually no more than a primary school education.[15] People from many different parts of northern Tanzania have been arrested, as well as many Meru. Sometimes rangers, looking to supplement their income, will kill park animals for sale in the villages; two rangers were arrested in August 1990 for attempting just that one night in Ngurdoto Village. The meat is then sold to certain butchers in the villages, who sell it clandestinely to local clients or mix it in with goat meat and beef. Ngurdoto Village is an area that has long been acknowledged as somewhat of a center for black market game. Sometimes game meat is used as barter and exchanged for livestock.

As for other resource crimes—fuelwood collection, grazing trespass, wood theft—it is, with the exception of the few outsiders living near the park, all local Meru peasants who are fined or arrested. The profile of the offender varies with the type of offense and is related to gender and age divisions of labor within the household. Women's crimes are generally restricted to fodder and fuelwood collection. Often they are apprehended in groups, such as the case where "8 women were caught and fined 180 Tsh,"[16] or another where several "women [were caught] who were crossing the park area going to draw water."[17] In the case of livestock trespass, it is the owner, nearly always an adult male, who is fined, regardless of who was herding at the time, which is sometimes a child.

The incidents of criminal activity also appear to vary with ecological conditions. It is, however, difficult to derive from park records firm empirical evidence to correlate seasonal changes with changes in the type and frequency of natural-resource crimes. There are too many other factors that influence the recording of violations—whether or not the ranger posts are staffed, whether or not the violations are *officially* reported (more on this in the next chapter), and whether or not the rangers are actually patrolling. As an example of this last point, in much of the area surrounding the park, April and May constitute the "hungry season" before the main harvest, and it seems reasonable to assume that this might be a time when hunting pressure would increase. The rangers, however, reported that heavy rains were keeping hunters and grazers out during the month of April.[18] The question arises as to how often and how diligently the rangers themselves were venturing forth to patrol in the inclement weather in order to make this determination.

Despite these difficulties, a distinguishable pattern of seasonal variation in violations of park laws can be found in the staff's reports. One report linked the frequency of violations to the agricultural calendar, claiming that hunting is low during periods when agricultural labor demands are high, but during the dry season, "when peasants are not engaged in farming," hunting is up.[19] Park administrators have also linked food scarcity due to drought to increased poaching. A warden once explained, "Sometimes the killing of animals increases in difficult situations. For example, between February and March, 1976 when there was a big drought, many buffalo and one giraffe were killed in different areas of the park. In the month of March, a total of 10 poachers were arrested. All of these were involved in killing buffalo and giraffe for the purpose of getting meat for food and for selling."[20] Perhaps more predictably, the seasonal changes in pasture availability and the occasional incidence of drought greatly influence livestock pressure on park forests and grasslands. The park warden's reports from a period of drought from 1974 complain of heavy livestock trespass, with six people arrested and fined in January alone.[21] In March the warden wrote, "There were continued grazing infringements into the park this month. Several cattle and sheep owners in the Kusare and Seneto areas were apprehended and fined . . . Generally, March was a heavy month for illegal grazing infringements. One cause was undoubtedly the extreme drought conditions all over Arusha Region."[22]

The other important management headache is patrolling and controlling the boundary against encroachment. The irregularity of the boundary, a result of the jigsaw puzzle approach to the park's establishment, and the land uses that surround it have made it difficult to regulate. Where the park is bordered by the forest reserve, people can come and go with relative ease under the cover of forest, and the Forestry and Beekeeping Division is limited in its desire and ability to offer assistance in patrolling. Where it is bordered by village lands, there is pressure to push the boundary inward, as in an area near Leguruki that was "illegally encroached upon by squatters[,] but from J. O. [John Owen, park warden in the mid-1960s] days[,] and I have been forced to accept it."[23] In another, unspecified, location the warden noticed that "the park boundary was shifted in by the villagers planting hedge seedlings."[24] The villagers often have a different view and complain that the rangers don't know where the real park boundary is and that people are sometimes wrongfully arrested as a result. While I was in the area in 1990, the park management demarcated the boundary by cutting a swath of brush, but villagers claimed that they were cutting several meters outside of the true boundary.[25]

TANAPA's response to persistent resistance from the Meru peasant communities surrounding Arusha National Park has been multifaceted, if lacking in consistency.[26] Much hope has been placed in a program of education, including the establishment of a mobile film unit designed to bring conservation education materials directly to the villages and rural schools. This particular education program eventually fell by the wayside as the projectors were eventually grounded for lack of spare parts. Attempts have also been made to solicit cooperation from local leaders: "At this time the action taken to reduce poaching was to visit and educate the ten cell leaders on the importance of conserving the environment by keeping the livestock from getting in the park and destroying it."[27] Along with their efforts to "educate" local peasants, park authorities have also pursued paramilitary-type approaches to park management. One former director was fond of reminding his charges that TANAPA was essentially a paramilitary organization (Bergin 1995). In line with this thinking, an Arusha park warden suggested in 1989 that·a "special unit comprising one or two platoons of soldiers should be formed to crack down on poachers."[28]

Historically, park administrators placed great emphasis on the expansion of the park's boundaries as the solution to its management problems. In 1974, for example, the park warden declared: "It becomes more and more apparent that every effort should be made to acquire the Momella Farm enclave which cuts right into the park on the Kusare Road between beacons TNN6 and TNP7. This enclave includes a large area of valuable wild olive trees, and allows the local residents to bring their stock within sight of Kusare House and very near our main road."[29] This area is currently claimed and occupied by residents of Nasula who use it as a forest commons. The management plan that did exist in the early 1980s pinned much of its hope for solving management problems on the expansion of the boundaries. Around Momella Lodge, for instance, a grazing commons used by Nasula Village residents was sought for inclusion in the park because "the proposed adjustment will therefore make it [the boundary] straight for an easy identification and control of human encroachment."[30] The plan suggested five other expansions, including the circumscription of all the natural forest in the forest reserve and other areas currently under cultivation.

Underdeveloping the Rural Economy

A popular argument in the conservationist literature in recent years is that parks are an integral part of rural development (see Neumann 1997a). Where jobs, sustainable access to resources, or revenue from park receipts are provided, a case can be made for this claim. For the villagers living in the immediate vicinity of Arusha National Park, however, the direct economic benefits have historically been few (cf. Bergin 1995). Tourism, for example, offers very little revenue beyond the occasional job as a porter for the infrequent climbing parties on Mount Meru. The Momella Game Lodge had only three local villagers employed in unskilled jobs in 1990. As part of park policy, local Meru are generally not hired to work as rangers in Arusha National Park for fear that they will cooperate with villagers. Aside from maintenance and clerical positions, then, the only jobs available in the park are temporary. When the park managers are engaged in a labor-intensive project such as road maintenance, they recruit casual labor gangs in the nearby villages—at the meager rate of 72.35 Tsh (about 36¢ per day) in 1990. Villagers are also allowed to use the park dispensary for free, and this is probably the most appreciated benefit for those villagers living close enough to the park's headquarters to take advantage. These benefits combined, however, are minor compared to the costs to the livelihood of people living and farming on the park boundary.

A number of authors have addressed the adverse effects of the establishment of national parks and game reserves on nearby rural economies in Tanzania. As detailed in the preceding chapter, parks and reserves were commonly created by relocating human settlement and activities outside of the boundaries. Kjekshus writes of the "de-development" (1977, 74) of areas of productive farm land that followed from the creation of the Selous Game Reserve. One result of relocation is the concentration of settlement and cultivation on the border of the park, leading to land degradation and the disruption of agricultural production by wildlife moving outside of the park's boundaries (Yeager and Miller 1986). In part, this problem is a result of the original boundary designations, made with little knowledge of the effect of human activities on controlling wildlife population numbers or of wildlife migration routes and habitat needs (Diehl 1985; Borner 1985).

One of the major themes of Kjekshus's (1977) influential but flawed work (Koponen 1988) on ecology control in precolonial Tanzania is that the creation of game reserves through resettlement greatly contributed to the expansion of the range of tsetse fly by eliminating the brush control measures instituted by African peoples, making vast areas unsuitable for livestock keeping. Iliffe (1979, 201-2) points out that, in the early colonial era, the creation of game reserves combined with human depopulation as a result of disease and European military conquest resulted in the spread of tsetse fly and the invasion of wildlife into previously cultivated areas. The documented adverse effects of Ngorongoro Conservation Area policies on resident Maasai livelihood included the halving of the number of cattle per capita, less milk available for consumption, and a decline in cash earnings (Arhem 1984). Finally, Homewood and Rodgers (1984) noted that the expansion of the wildebeest population as a result of their protection in the Ngorongoro Conservation Area has directly conflicted with Maasai livestock herding.

The productive activities of Meru peasant households near Arusha National Park are hampered in a number of ways because of the proximity of wildlife populations. From the villagers' perspective, the most critical aspect of the management conflict is the destruction of food crops by wildlife coming from inside the park, for which no compensation is offered from the government. For those people who are unfortunate enough to have cultivation plots on or very near the boundary, the destruction is often complete, sometimes occurring overnight. A farmer on the boundary in Ngongongare explained, "If you take care and guard your crops, from one acre planted you can harvest a quarter of an acre."[31] Walking along the park boundary, one can observe maize plots that are 50 to 100 percent destroyed by elephant and buffalo (see figure 9). A Nasula resident similarly complained, "If I plant seven acres

Figure 9. An Ngongongare resident surveys the remnants of his maize and banana crop, destroyed overnight by park wildlife. (R. P. Neumann.)

and guard it without getting any sleep I'll get three and a half acres."[32] Even permanent crops, such as banana, were subject to destruction by elephants. One group of brothers in Nasula whose inherited plots happened to end up on the boundary after one of the park expansions were not even attempting to plant crops there. Most people do not have the option, as these brothers did, of moving production to another plot of land out of harm's way.

To further exacerbate the problem, the local agricultural calendar and the seasonal movements of park wildlife express an unfortunate convergence. The buffalo and elephant move seasonally within the park, migrating to the lower elevations during the height of the rainy season. A twenty-nine-month study of crop raiding by elephants from April 1966 to August 1968 by Vesey-FitzGerald showed that raiding peaked following a period of heavy rainfall. This means that the bulk of the elephant population is in close proximity to the farms at precisely the time when the maize is approaching maturity. Bush pigs and baboons, as well as numerous other bird and mammal species, are also

a constant source of destruction, though perhaps not so absolutely devastating as larger animals.

Both villagers and park management have attempted numerous
experiments to minimize these conflicts. In the early years of the park,
a serious effort on the part of the administration was made to reduce
destruction and research was conducted to discover patterns of crop
raiding and even to identify individual "problem" animals. In 1966, the
park management installed two electric fences, one 9.6 kilometers long
near Leguruki and another 2.4 kilometers long near Lendoiya. The idea
was to create a "fear barrier" rather than a physical barrier.[33] Neither
was completely effective (bull elephants soon learned to short-circuit
the wires),[34] and they required intensive maintenance that could not be
kept up within the context of Tanzania's economic problems and the
insufficient funding for parks. In recent years, there has been no strategy formulated or initiative of any kind by park authorities to control
the movement of park wildlife into the surrounding farms. There have
been sporadic interventions by game scouts in serious cases, such as in
1988 when they killed a buffalo that had invaded a farm and threatened
the owner.[35]

In even more infrequent cases, park rangers have attempted to
defend villagers' plots against wildlife,[36] but the policy of the park de-
emphasizes this as a normal duty. In fact, the park authorities have
worked to also reduce the role of Game Division scouts, even though
historically one of their important activities has been the elimination of
wildlife that were damaging crops or killing livestock. As early as 1966,
an informal arrangement had been made between the Game Division
and the park not to shoot park wildlife that had strayed into surrounding farms. The regional game management officer acknowledged the
agreement:

Senior Game Scout Saidi Kawawa of Usa River has told me that you
informed him that the park personnel would undertake the chasing of game
such as elephant, buffalo and rhino from surrounding the park. This was to
ensure that animals which strayed from the park did not get shot during
crop protection.

    Would you please make an effort to inform farm owners about the deci-
sion[,] because most people keep on complaining to the Game Division[,]
saying that the Division is not adequately protecting the farms.[37]

This policy has been partly based on a fear that wild animals would
become so frightened of humans that they would remain deep in the

park, out of sight of the tourists. The warden wrote in reply to the above, "But the animals should not be shot at all because the *shambas* [cultivated fields] are quite close to the park[,] and if we allow shooting in this area all the elephants will move right into the forest where they will be seen by nobody."[38] Since some species have been extirpated from the park, a stronger ecological rationale is used today. "Shall we go on killing when these rare species are disappearing?" the warden wrote in 1989 in a letter requesting the Game Division not to shoot park animals in farms.[39] The current policy at Arusha National Park is based on an agreement between the warden and the local Game Division office to not shoot any animals in the crop areas. Consequently, local farmers are left nearly defenseless in the face of raiding animals.

Villagers have, however, organized themselves to dig trenches in several locations, and these have occasionally been effective when properly constructed and located. It is not an entirely workable solution, though, because of the stony soils in many places and the constant demands on labor to keep them cleaned out. In addition, the villagers' efforts have often sparked boundary disputes with the park authorities. When the villagers of Seneto took it upon themselves to dig a trench and plant a living fence, they were chastised by the warden (who feared encroachment) for not communicating their plans to the park before work began. He ordered them to bury the trench, pull up the trees, and pay for the petrol he used to visit the site of the activities.[40] A week later, the villagers had not complied and the warden threatened the village chairman with legal action.[41] He instructed the villagers to dig the trench not on the boundary, but three meters away from the boundary *into* village land.

A practice among villagers in Nasula is to build temporary grass huts in the middle of the cultivation plots and sleep there during the season while the crops are maturing, hoping to chase the wildlife back into the park. Since very few people can afford to buy guns, or even bullets, which might be used to frighten animals, banging pots or waving or flinging flaming sticks are common ways people attempt to spook elephants and buffalo away from their crops. This method is practiced in relatively few plots as it is difficult to get access to enough grass (or other materials for that matter) to build the hut, and the results achieved often do not justify the hazards of facing down a buffalo or elephant in the middle of the night. The desperate character of this "solution" is an indication of the degree to which villagers' ability to meet their subsistence needs is threatened by marauding wildlife.

These factors have facilitated an ingenious (for *Meru* farmers) solution to the crop-raiding problem and an interesting pattern of land ownership. Since Ngongongare is close to the main tarmac road and fairly close to the town of Arusha, it is not uncommon to have non-Meru settled there. Much of the village land, having been settled only after the European estate owners left in 1969, is not subject to customary law and can be subdivided and sold. All the farms along the park boundary have been subdivided and sold to *non*-Meru (several are Pare and Chagga), forming a de facto buffer zone for Meru farmers. In the bitter summation of a farmer from distant Musoma, now living on the park boundary, "We form the barrier for the Meru. The animals eat their fill in my farm and *mzee* there sleeps soundly."[42]

Infectious diseases are also transmitted between livestock and wildlife. Rinderpest, as well as east coast fever and trypanosomiasis, the eradication of which is difficult where there are concentrations of wildlife (Mugera n.d.), is a problem in villages to the north of the park. Veterinary services are limited in the Ngare Nanyuki Ward, and this fact combined with the proximity to wildlife makes it virtually impossible to raise the more productive but less hardy cattle breeds.

People living on the park boundary and going into the bush of the village commons to collect fuelwood or graze cattle are under a daily threat of personal injury and death caused by wildlife. The correspondence of a former park warden illustrates the severity of the problem: "As a matter of interest, in July alone, in the never ending clash between game and humans, two buffalo were wounded and one young elephant killed by settlers in the Lendoiya section and one settler was killed by an elephant in the Londuka area."[43] Women generally will not venture into the bush alone, and because of the danger their husbands occasionally assist them in gathering fuelwood. During the period of my research, at least three people from villages on the boundary were killed by park wildlife and several others were injured. In October of 1989 a buffalo gored an adolescent boy herding cattle in the bush near the park boundary in Nasula, and he died later in the hospital. Around the same time a family of three was attacked in the village, but all survived with rather vicious injuries. Buffalo killed two more people in Nkoasenga Village on the other side of the park several months later.

Predators coming from the park can wreak havoc on livestock herds literally overnight. As an example, one afternoon in Nasula in early March a six-year-old child was left alone by his older siblings to watch over a herd of his grandfather's sheep. As evening fell and his brothers did not return, the boy abandoned the herd and returned to his home without informing anyone about the sheep. That night, thirty out of fifty sheep were killed by hyenas. The next morning a dozen men gathered to butcher what was left to try to salvage something from the loss, but the owner only shook his head and said, "[M]y herd is finished." This is an extreme incident of livestock deprivation, but not an uncommon one for people living near the park boundary, who must remain constantly vigilant.

6

Village Moral Economy
and the New Colonialism

Contested Myths and Meanings

Many days while I was talking with villagers in Nasula, Yesuria was off working in his fields or on other business. In the late afternoons it was our habit to come back to Yesuria's homestead and sit together in his mud-walled house, drinking tea or coffee. I would use him as a sounding board for my ideas and interpretations and he would talk about local history, the affairs of the village, and the park. Looking out the open door over the fields and pastures not so long ago occupied by European settlers, he would tell me about life under British rule. One particular story, about a settler named Monas who once held the land where we sat, is revelatory.

Monas was very fierce. He charged money for people to graze on his farm and if we would go to him to ask for meat he would say, "Don't even ask. Meat is for white people." Well, when Monas wasn't around, the Meru who worked on his farm would look for a nice fat healthy cow. We'd get a bit of water boiling and force it down that cow by holding its nose. Within half an hour it would be dead. Then we'd go tell Monas, "Mzee, one of your cows has died." And he'd come over to have a look and say, "It's anthrax. Don't take any of the meat. Don't even let the dogs get near it. Just dig a hole and bury it." We'd bury the cow in a very shallow hole and then come back in the evening and butcher it right there, bury the skin, and take the meat home. And after we'd eaten the meat, we'd bury the bones. That European would say, "Oh, I've got very good people. I can trust them totally."

I recount this story because it illustrates nicely the central arguments of this chapter and indeed much of the book. It is an exemplary tale of how local patterns of resistance manifested themselves in the daily life of Meru peasant society. This act of theft is typical of the forms of resistance that first emerged under colonialism and continue today. Characteristically, the act had to be planned and executed so as to remain hidden and undetected from "above," because of the inherent danger to the participants should their actions become known. As with all "crimes," the most effective theft is one of which only the perpetrators are aware. Yet, at the same time, there was clearly widespread knowledge of and community participation in the crime and a sharing of the "booty." This was not cattle theft for individual gain, but an effort to redistribute wealth in line with a local normative standard.

Second, it gives us a glimpse into the culture of a subordinated people, within which theft and deception are locally celebrated and morally justified—celebrated in the sense that the story was told in the fashion of an inside joke, the punch line being Monas's misplaced trust in his workers. And, it was morally justified in the sense that one hears in the telling of this tale an implicit claim of injustice done to Meru farmers for having to pay to graze on their own land, and of the wrongness of the estate owner who denied requests for food. It reveals local expectations of a system of mutual obligation, however unequal (Polanyi 1957; Scott 1976), between workers and the estate owner. The theft of the cow was not intended to radically challenge the social and economic position of the Meru squatters vis-à-vis Monas nor to drive him from the land. It was a direct attempt to gain justice after having failed to convince Monas of his moral duty to those much less wealthy and powerful than he.

The interpretation of the story is incomplete, however, if we leave it as simply a demonstration of the tensions in patron-client relations and normative ideas of mutual obligation. For in this particular case, the would-be patron is not only a large landholder, but a usurper of local land claims. The moral justification of certain crimes, such as theft and grazing trespass, are to a large degree based on claims of ancestral occupation and customary rights to land and resources. Simultaneously and inseparably, these acts defend a set of locally recognized symbols and an idealized history in the land—an ancestor's homestead, a meeting place, a battle site, a communal salt lick. It would be misleading, however, to imply a homogenized "Meru" identity and history embedded in the local landscape. There are, instead, many (sometimes competing) histories and overlapping identities, some quite restrictive, others more inclusive.

Much of the upper reaches of Arusha National Park, for example, is claimed as the ancestral lands of two of the larger clans, Nnko and Mbise, that comprise the Meru. Sori Kibata is a ritual leader for the Mbise clan. As the bearer of knowledge of his founding ancestor and his clan's history, he reads the contours of the mountain like an archivist. He offered to share some of this history with me, and for three days I followed him through the landscape of the mountain's western flank. As he wove a path marked by signs illegible to me, more than once I found myself clawing up particularly precipitous passages on my hands and knees. At one point he explained, "Many names in the park are from the people who lived there before. If they didn't live there, then they visited in that area a long time ago and did something there." From a vantage point on the crater wall, he swung his finger across the forested slopes below. "From Njeku going down, you have the place of Lamirei. His son, Chara, stayed at Kitoto, and further down from him lived another son, Mangala, below Pilo Hill. Where Campsites One and Two are now, there was a person called Nassary. In the area of Campsites One, Two, and Three, there lived a mixture of clans, including Nassary and Mbise." As we moved southward through the forest, he named the small tributaries and other features of the landscape, names not to be found on official maps. "Below, there, where the Pilo and Wato Rivers come together, we call Sangananu [where they meet] . . . We call this clearing Nrwai [a thorny tree]. Before, when coming from Meru along this path, we had to stop to discuss our strategy for crossing the clearing because the area was full of rhino."

These passages point toward a unity of social identity, local history, and landscape. This locally lived geography is as much an expression of identity as a declaration of rightful possession. Lineage provides not only one plane of a multidimensional social identity, but a set of rights to land and resources that has been progressively eroded by state conservation policies.[1] Resistance can be understood not only as an effort to maintain claims of access to subsistence and other perceived rights. On Mount Meru resistance to state-imposed conservation policies is also about defending locally constituted meanings etched in the landscape. The current conflict between the state and Meru peasants is both a struggle over land and natural resource access and a contestation over landscape meaning and representation.

The European settlers are now gone. Significant portions of their former estates lie not in the hands of Meru farmers, but behind the boundaries of the national park. The land has taken on new meanings derived from European representations of Africa. The Western national park ideal is by now familiar, but it is worth a short digression into travel writing on Mount Meru to reinforce the idea of European-derived meanings in juxtaposition to the above passages. The late poet and author Evelyn Ames was much taken by Arusha National Park, describing her experience there as, not surprisingly, like being "alone in Eden" (1967, 146). In her account of leaving the park we can hear many of the themes of nature that African national parks were meant to embody for Europeans: the park is primordial, undisturbed, unchanging, and pure in the absence of humans. Included in her anecdote is, significantly, the distinct separation of Eden from the world of human affairs: "Finally, he ducked down into the tunnel of green and we moved on down our own larger tunnel and out the park entrance where in the deepening dusk a dark figure stepped forward, saluting us, and the gate swung shut behind. The gate of Eden it felt like—with all that diversity and wholeness and harmony shut away now behind us in the night, the dance of eating and being eaten, creating and dying continuing undisturbed, as it has been since the beginning, was now but might not be forever" (Ames 1967, 151). The representation of Arusha as a prehuman remnant providing refuge from society is also developed in another popular depiction, where the park provides "a sense of a complete withdrawal from the world of man and of immersion in the peace of unspoilt nature" (Vesey-FitzGerald 1967, 13).

Tanzania's independent government has accepted the national park model based on these Western ideals of pristine nature. Arusha National Park remains principally an attraction for foreign tourists to experience "primeval Africa." It should be clear from my discussion of their history that for some Meru the park is imbued with very different meanings. To begin, we find two contending origin myths. The Judeo-Christian creation myth of Eden, albeit greatly abstracted and secularized, provides a convenient representation of Africa and, particularly, the national park as some pure place that Europe has lost (Anderson and Grove 1987). This reference essentially appropriates the landscape for Europeans by presenting it in an understandable and familiar idiom that simultaneously negates local stories of origin. The Meru clans' mythologized histories of their founders, in contrast to the abstracted references to Eden,

are very concrete representations of a specific place.[2] Contained within
the oral remembrances of lineage founders' lives and deeds is a pre-
sumed legitimation of claims to rights of access to land and resources.

Furthermore, the national park authorities deliver a message to Meru
peasants strikingly similar to the one delivered by Monas more than
three decades ago. The national park, like the settler's beef, is for white
people. While the message is not so bluntly stated, it is implied in the
government's insistence that the landscape of Mount Meru is a source
of *national* revenue—a landscape for cash-bearing foreign tourists to
consume. Meru representations of the mountain stand in opposition to
a landscape of consumption for tourists. The mountain stands not as a
landscape of production in the capitalist sense, but as part of the means
of reproduction for the Meru and a source of their identity. It not only
represents their history and their links with ancestral kin, but the mountain landscape itself is the *physical manifestation* of their history. Furthermore, the resources found on the mountain were also key to their
material well-being as a source of dry-season pasture, fuelwood, and
building materials. There is, in sum, no landscape dualism in Meru peasant society that corresponds to the landscapes of consumption, landscapes of production, or the insider/outsider dichotomy that we find in
the cultural production of nature under capitalism.

The story of the cow theft serves to reemphasize the continuity of
the contemporary situation with the historical relationship between
Meru peasants and the colonial state and European settlers. The continuity of the relationship, as will become clear below, manifests itself
multidimensionally but still revolves around the control of access to
locally claimed lands by a powerful outside entity. It is of no minor significance, for instance, that what were once alienated estates, with all
their symbolic baggage of colonial injustices, violence, and deprivation,
are now part of the national park. The designation of settler estates as
park constitutes yet another layer of meaning spread across the mountain landscape, one that cannot be untangled from those that were laid
before.

There were, to avoid crude generalizations, possibilities for more
multidimensional and less antagonistic relationships between settlers
and Meru peasants than implied in the story. Specifically, the story hints
at the potential for a paternalistic relationship between large landowner
and his peasant laborers, a space (however small) for negotiation over
the rights and entitlements of the poor. Additionally, as we have seen in
chapter 2, there was some advantage to working on estates, particularly

access to new knowledge regarding the farming of cash crops. There also existed a settler minority who, though unwilling to give up acquired lands, were open to peaceful coexistence instead of the continued mass evictions of Africans.

In sum, the historical relationship was dynamic and contingent, shifting between resistance, negotiation, and accommodation, all the while viewed by Meru peasants through the lens of a local moral economy. The national park authority now controls access to lands previously controlled by European settlers and the colonial state. The relationships between the park and surrounding Meru villages, I argue, share some common characteristics with those historical relationships. Contemporary acts of resistance and the interpretations of morality and justice revealed in the discourses of Meru peasants resonate strongly with their experiences under colonialism. The theme of moral economy is woven throughout this chapter in multiple strands. The local moral economy, anchored in the right to subsistence and customary land claims, structures the meanings of justice and fairness. In this chapter, I will demonstrate that park laws and the crimes that violate them are measured against the normative standard of the local moral economy. Laws and policies that violate it are resisted through word and deed. Thus some park crimes are viewed locally as morally justifiable.

"The Citizens Are Begging Very Much"

Colonial administrators, as we have seen, often expressed their fear of local rebellion over the loss of customary rights on Mount Meru. There have been, however, no actions resembling anything like a violent revolt on Meru since 1896. In no small measure this is due to the Meru's experiences with the power of state violence, first under the Germans in central Meru and later the British at Ngare Nanyuki. Villagers neighboring Arusha National Park often spoke of the weakness of their position vis-à-vis the state. It is fair to paraphrase the common sentiment as: "The government is strong, it can do as it likes." Rather than violent confrontation over their loss of access rights to the park, the response of Meru villagers has been similar in character to their response to the British eviction. More generally, their response parallels the activities of the preindependence nationalist movement in Tanzania in general (Iliffe 1979). That is, they have sought to have the perceived wrongs of the colonial and postcolonial states rectified by petitions and appeals through established official channels.

This has been particularly evident during times of dearth. For instance, during a 1982 drought, grazing pressure on park lands intensified, prompting the warden to contact the Seneto Village chairman. The warden, like so many of his predecessors and Forest Department counterparts, requested the chairman's assistance in stopping cattle from entering the park. The warden's appeal had the opposite of its intended effect. Rather than slowing grazing trespass, the request sparked a village protest over the closure of access to park resources. The chairman described the incident in a letter to the TANAPA director: "This office has received orders from the Chief Park Warden, Momella [actually, Arusha] National Park[,] that it is prohibited for villagers to pass their herds in a corner of the boundary in the Seneto section of the park. After these orders, the villagers arrived at the office in a procession[,] countering that where should they pass their herds in order to drink water in Seneto Pool, obliging the office to contact the chief warden on 2/10/82. After long debate, the park warden said that he himself was unable to give a verdict which does not come from the Director, but he allowed their cattle to pass for a period of just two weeks beginning 3/10/82."[3]

In discussions and correspondence with the warden, and ultimately the director, the chairman petitioned the state to recognize rights of access that had been denied since the path's closure. The villagers were in essence attempting to defend their rights to subsistence. The village chairman proceeded in his letter to the director to point out that the village herds are dependent on Seneto Pool as a water source, and have been for many years: "This is the customary path from long ago[,] which was passing through Seneto up to Meru as it comes from Ngare Nanyuki to Meru. For the past 9 years we have been shut off from this path and we are absolutely suffering." He closes his letter with a request to the government to reconsider the closure. "Even though the law is understood[,] this office together with all the citizens are begging very much that we should be considered for passing this way." The tone of the letters is deferential, and they contain an explicit recognition of the authority of the government to enforce park laws. They make, instead, a moral appeal to government officials to recognize the injustice of the laws' enforcement during the drought and to consider the validity of local customary claims. The park authorities turned down the request, presumably to avoid setting a precedent.

The path to which the chairman was referring to was, of course, the three-meter "exclusion" through the park, which passes by Seneto Pool, historically an important watering hole. I visited the chairman one September afternoon in 1990 near his village office. He became thoughtful when I asked about the closure of the right-of-way, and it was clear that the issue was of particular concern to him. The chairman took great pains to explain its history and the numerous steps he had taken through official channels to get the injustice redressed: "After the park started in 1963, we were still using this path. We had no problems crossing the park and there were still plenty of [wild] animals. Then in 1973 or '74 the path was closed. There was no official communication, it was just closed by force and we failed to get anyone to represent our case."[4] The closure of the path has had, according to the chairman, a significant adverse impact on the villagers' resource base. Through his positions in Chama Cha Mapanduzi (CCM, the national political party) and local government, he raised the issue when he became chairman in 1982: "We followed all of the channels from the district to the Speaker of Parliament, Adam Sapi. The Speaker wrote to Babu [then TANAPA director] and asked him to reopen the path. Babu has done nothing. I wrote another letter asking him, still nothing is done. This was in 1985 and '86." Despite his continued petitions to state authorities to correct the perceived wrong of the national park officials, no action had been taken by any state agency by late 1990. The chairman had not given up, however, and told me he was planning to renew his letter-writing campaign once the 1990 national elections were completed.

During an earlier drought in 1974, village leaders had made similar formal appeals to officials to help them with the scarcity of grazing. Livestock herds had come under stress since the high-elevation pastures within the park's boundaries, once so important to the local grazing regime as drought reserves, were now off-limits. With little alternative, farmers attempted to bring their starving herds into the park. The warden's reports from that period complain of heavy livestock trespass, with six people arrested and fined in January alone.[5] Their subsistence limits strained, local residents again sought relief through official channels, asking for restoration of their rights of access. On February 4, the warden was invited to a village meeting. He was formally requested to allow residents to graze their herds in the park due to the drought. As the warden noted afterward, "Their request was turned down on the spot because it was contrary to the National Park Ordinance."[6] Despite this cursory rejection, park officials responded favorably to a group of

Meru elders who, in an effort to end the drought, requested permission to perform a ceremony at the ritual site at Njeku. A heavy and prolonged rain, though delayed a week, followed their prayers.[7]

If there is any lesson that Meru farmers have learned from the incidents recounted above, it is that the ears of the bureaucrat are apparently deaf to arguments for flexibility in the enforcement of national park laws. It is also clear that claims of access to land and resources, when backed up by nothing more than oral histories and local custom, carry little weight with officials. Access rights, if they are to be defended, have to be demonstrated in ways that the modern bureaucrat recognizes.

Encroachment or Tenure Security?

When these appeals fail, it can only serve to remind neighboring villagers how tenuous a hold they have on land and resource access. Historically, colonial governments defined African land rights quite narrowly in terms of occupation (Coulson 1971). Land that was settled or was under active cultivation was generally considered occupied, while most forests and much seasonal or drought reserve pasture, where there was no permanent settlement, was declared public land. In colonial Tanzania, the limited recognition of existing claims was codified in the 1923 Land Ordinance as "rights of occupancy" (James 1971), and this was maintained after independence. According to a 1991 presidential commission in Tanzania, the most substantial change from colonial to postcolonial land law (until the government initiated a recent land-titling project) was to substitute the word "President" for "Governor" (United Republic of Tanzania 1992). The legal situation thus makes security of tenure for lands not under cultivation (pasture, bush, forests) extremely vulnerable, particularly to claims made by the state (Bruce 1993, 50–51).

The repeated failures to elicit any support from conservation authorities for restoration of lost rights have made villagers all the more aware of the need to strengthen their claims on the land and resources to which they still hold recognized rights. In the words of one ten-cell leader, "If we weren't brave, we would lose our land."[8] For instance, ownership of and access to a strip of the former Momella Farm has been a point of friction between the park administrators and local residents since the park's inception. Nasula residents recognize that their claims are vulnerable, as they have watched the park boundaries move progressively outward. They fear, and records support their fear, that park authorities covet the village's land, and so villagers have launched several collective initiatives designed to strengthen their claims to the area. The most controversial has been the building of a schoolhouse in a plot taken out of the village's grazing commons, lying between the village houses and the park boundary.

As related by one of the original residents, the action was a calculated political tactic, inspired by a former owner of nearby Momella Game Lodge. He explained that "Mallory [the former lodge owner] guarded us, guided us, and gave us the tip of building a school, of which we had no idea . . . [W]e live here as our permanent home now."[9] This was an effective and brilliant move on the villagers' part, both symbolically and materially. Symbolically, because of the importance that former President Nyerere and his government had placed on education infrastructure in rural areas and local initiative. The negative symbolism of government officials trying to remove or destroy a schoolhouse would cause park authorities to think twice before attempting to take this land. Materially it was important because evidence of occupancy and use is key to gaining official recognition of ownership. The school helped convert a former squatter settlement into a permanent village. In addition, if the village loses and must be relocated, claims for compensation will be far greater.

Concern over compensation as well as a desire to increase security of tenure compelled villagers to act on another piece of grazing commons. Around the time the school was built, Nasula residents were also formalizing claims to land on the park boundary. As part of the project to receive recognition as an administrative unit (*kitongoji*), village leaders requested the district government to recognize the subdivision of the commons into individually held plots. Land registration would of course also greatly strengthen ownership claims. The importance the villagers place on this strategy was demonstrated in mid-1989. At that time, a rumor circulated in Nasula that the government intended to extend the park boundary to encompass village lands. A meeting of the village committee was quickly called and a collection taken up to provide funds for a delegation of three villagers to go to the district office to clarify the rumor. The delegation returned with no clear or satisfactory answer. In a second meeting the village committee members "agreed that in every plot where someone is living they should plant permanent crops, and those who have not moved there should do so at once."[10] Again, the move is an explicitly political initiative aimed at strengthening the village's position vis-à-vis the state.

These actions can be understood as conscious strategies to strengthen land claims in a climate of uncertainty of tenure. But the notion that "encroachment" of village settlement on the boundary is in part a response to the atmosphere of insecurity of tenure produced by conservation policies does not seem to have occurred to officials. For park administrators, the root causes of these "threats" include criminal intent, population growth, and a lack of understanding of conservation by local residents.

State Crimes, Local Justice

In between the periodic petitions to officials for the recognition of lost rights and park access, the women and men of neighboring villages labor daily to meet their needs for food, fuel, and shelter. The lands inside the park boundaries have long been one important source in the struggle for subsistence. With the general failure of these petitions to resolve the problem of resource access, villagers have continued to rely on the tactics of deception and secrecy developed under colonial rule. Some villagers boasted about being able to come and go as they pleased without getting caught because they know the park so much better than the rangers from outside the area. However, others felt that it was becoming more difficult to circumvent the park's law enforcement structure. People who used to regularly go into the park to collect fuelwood or graze their cattle are now hesitant to enter for fear of facing arrest. Sitting in his house, the park boundary in view, Sarkikiya told me, "Before we could take our cattle into the park late on Friday or on Sunday because we knew there would be no one around. Now there are so many guards that it is difficult to graze our cattle in the park."[11] Others voiced similar frustrations about timber collection: "Years ago it was good because there was only one guard in the whole area from Meru to Ngare Nanyuki. Now even getting into the forest we are afraid to cut timber for building because we may cross the park boundary and get caught."[12]

Commonly, people still try to adjust the normal time and place of resource use in order to avoid detection. Villagers sometimes try to take their cattle in late in the evening or collect fuelwood and building poles at night. The warden's monthly report once noted "that most of the destruction takes place during the evenings. For example, one day four people with livestock were seen in the park at 7:30 P.M., which is not normal hours."[13] The ends to which villagers will go is demonstrated in a case where rangers arrested someone cutting trees in the park at 2:00 A.M.[14] Understandably, few villagers would readily discuss the details of how they stole wood in the park. Sorana, a farmer living in one of the villages near the park was, nonetheless, eager to demonstrate to me how easy it was to enter the park undetected. Villagers essentially use the cover of the surrounding forest in the evening hours to steal into the park without being observed. Passing through the forest reserve where there are relatively few guards, villagers can obtain brown olive (*Olea africana*), a species highly valued as a fuel source, which, in some areas, is now found only in the park. Of course, park officials are well aware of this and it is part of the reason they had long sought to include all of the forest reserve within the boundaries of the park. Ultimately, all the parties involved acknowledge that government control over the park has tightened in recent years, making even the tactics of secrecy less and less viable.

In the course of my observations and conversations in the village, and in my research of park reports and historical records, it became clear that the social and ecological threads that bind together the national park and local residents are woven through time, providing the conflict with historical continuity. Through British rule up to the present, this continuity is expressed in the professional reports and in the reaction of residents to state laws governing the use of natural resources. Exasperated by their failure to halt grazing in the Meru Forest Reserve, colonial foresters placed the blame at the feet of local authorities. "It would seem reasonable to suppose," wrote the conservator of forests in 1928, "that the reluctance shown by the Native Court, which is in sympathy with the offenders, to take action has had some effect in the great increase in this form of offense."[15] More than a half century later, there is a revealing similarity in the report of the Arusha National Park staff, who complained that "the cooperation with the villagers is not so good because when poaching activities are reported to the village leaders, they promise to deal with them[,] but as a result nothing is being done."[16]

With the establishment of Arusha National Park after independence, and the subsequent elimination of all customary rights within its boundaries, local resistance to state conservation policies has continued much as it did under the colonial government. While the local Meru courts are no longer responsible for prosecuting violations of natural resource laws, village chairmen and ten-cell leaders, as representatives of the state and the ruling party, are responsible for cooperating with their enforcement. However, just like the Native Courts in colonial times, the park is receiving little support from them. The park reports complain of a lack of cooperation from village leaders, and one warden wrote, "It is very discouraging for the park staff to hear that the ten-cell leader is the one who allows nearby villagers to kill buffaloes."[17]

These local-level officials are, particularly in the more subsistence-oriented villages, principally peasant farmers themselves. Kilakikya had been a ten-cell leader for over a decade when I met him in 1989. Sitting in a torn, patched shirt in his mud-walled house, he made no effort to disguise his bitterness toward the park. He personally had been fined four times in recent years for allowing his livestock to trespass in the park. He was, in fact, one of the most outspoken critics of the park's policies I met, especially on the issue of crop damage by wild animals. I had often noticed when passing his house a semicircular arrangement of the skulls of four buffalo (the scourge of farmers) near the front door. When I pointed them out, he remained placid and replied offhandedly, "I found them in the bush."

In another village on the southern boundary of the park, I sat one afternoon with a group of elders, listening to their complaints of past injustices done in the name of wildlife conservation. A story they related clearly illustrates a solidarity within the village, including the leadership, with regard to a local understanding of criminality. One of the elders recalled,

One day in 1983, I found a bush pig in my farm eating the maize. I speared it and took it home and had the head hanging outside. The park guards discovered it and five of them came to my house and told me to get the head and carry it to the police. I refused and told them I wouldn't take it unless they carried the maize which the pig had been eating. This was the real offense and the maize was the evidence. We argued for a while, and then the guards sent for the ten-cell leader for help. The ten-cell leader also told them to go and collect the maize for evidence. Then they went to the village chairman and he told them the same thing: Go and take the maize. After this the guards gave up and went back to the park.[18]

The solidarity of local leaders with the farmer's interpretation of justice is common to moral economies of the poor in other historical periods and places (Thompson 1971; Scott 1986). To elaborate, it is reminiscent of Thompson's eighteenth-century mobs "rescuing" arrested rioters from the authorities (1971, 121). The story expresses moral indignation at the conservation laws of the state that allow farmers' subsistence security (Scott 1976, 5) to be threatened. The "crime" of poaching is thus not a crime at all, but a defense of subsistence, and the "real crime" is that park animals are allowed to raid crops with impunity. To be clear, I am not arguing for a relativist analysis of crime and justice. Rather, I am assuming "that household heads wish to safely ensure adequate consumption streams, that is, simple reproduction" (Watts 1983, 16), and this wish is the basis for the formulation of local ideas of criminality.

Today it is difficult to find a household in Nasula that has not had to pay a fine for a park violation. In Ngongongare, a resident responded to my inquiries about how much he had paid in fines, "In my case I can't tell you because I've been fined almost every year. Now which year can I tell you about?"[19] Many of the people in leadership positions within the villages, from church to ruling party, have been fined for grazing cattle in the park. The lamentations of successive generations of natural resource professionals indicate that the local leadership, if not directly undermining the policies, is apathetic at best. Though some of the local response to the criminalization of customary rights is characterized by individualized acts, these occur within an environment of community acquiescence, where park violations go unreported, the identities of the lawbreakers are protected, and village leaders withhold their cooperation in solidarity with their constituents. Theft and trespass cannot be categorized as the isolated acts of social malcontents, for they are part of a pattern of village opposition to state policies that violate the local moral economy. It is these violations by the state and its agents that have produced the community-based pattern of resistance to national park policies.

Violations of Village Moral Economy

Listening to the voices of Meru villagers who live and farm on the edge of Arusha National Park, we can learn something of the essence of local resistance to park policies, of local standards of justice, and about the ways in which state policies and officials are seen to violate these standards. The most obvious and explicit violations result from laws and regulations eliminating customary rights of access in the national park. For instance, villagers who claimed customary tenure to beekeeping areas in the park, with as many as twenty-eight hives per person, claimed that their hives were removed from the forest without consultation or compensation. An Ngongongare farmer recalled, "People were keeping very many hives, and their grandfathers would go for even a week to collect honey. But after the national park everything has been prohibited; even a person found wandering in the park will be fined."[20] Another farmer, when asked how the park had affected the village, replied, "We haven't faced a problem with cultivation, but for grazing it has been a setback, since the park has taken a lot of resources from us."[21] A woman who had lived in the Momella area since 1950 remembered customary practices before the park: "We used to get ngyesi, sakyona, usokonoi [common medicinal herbs]. Also we collected firewood and cut timber for building our houses at that time. We don't get this nowadays, since the park has protected all of the area we used to enter freely before."[22] In response to my inquiries, Meru residents emphatically claimed the national park as part of their lands. I found this claim invariable throughout Meru villages, though its significance for local land use varied greatly. To elaborate, it was among the villagers living in communities near the boundaries where I found people with the strongest memories of the park lands as a place for grazing, resource collection, settlement, and ritual. Even within these villages, intergenerational memories differed significantly. That is, many of the elders remember living as squatters on lands that later became national park, or remember collecting in the forest. The younger generations may remember grazing or collecting fuelwood after the park was established. One farmer, who was barely a teenager during the park's early years, remembered that "formerly there was one park manager who allowed firewood collecting and erected a fence [to reduce crop raiding by wildlife]. That helped a lot, but now they don't allow it. A long time ago Vesey-FitzGerald [a former park scientist] was helping us a lot; because the park was small, he could look out for all sides."[23] Under these circumstances, villagers recall that access rights were recognized through agreements with park wardens and guards.

The point is that the basis for claims of customary rights and practices is highly variable within village society, ranging from memories of long-ago personal experiences to stories of ancestors heard from one's grandparents. Nonetheless, as losers to the park in the struggle over land, virtually all villagers remember the "good old days" (see Scott 1986, 148–51) when customary rights of access were not so thoroughly challenged. In the absence of title deeds, the memories of historical land use practices and ancestral settlements serve as the principal legitimation of local claims. These reminiscences are thus a central element in the struggle with the state over land and resource rights. On Mount Meru,

as in much of Africa, "the power to interpret—or reinterpret—historical events in order to legitimate claims to land is often of critical importance" (Shipton and Goheen 1992, 309).

Local efforts to have customary claims recognized, however, are not
based solely on backward-looking appeals to past practices. Many economic and cultural changes have swept through Meru society over the
past century, and Meru efforts to legitimize their claims sometimes
reflect those changes. An important example is the case of the ritual site
for the Mbise and Nnko clans on the floor of the Mount Meru crater.
A ritual leader of the Mbise clan complained that it was difficult to find
enough people for the ceremony because most people, including himself, had now been baptized as Christians. The ceremony, historically
held around January to appeal to the ancestors for rain, requires the participation of non-Christian boys and girls, who are becoming increasingly rare. Kibata explained that "these days it [finding nonbaptized
children] is a bit of a problem, so we want to ask to go to the park and
pray like Christians."[24] He plans to ask permission to build a church at
Njeku (the ritual site) and continue going to the mountain with just the
Mbise and Nnko clans, still taking a cow for sacrifice. In the context of
cultural change, local tradition is reinvented (Ranger 1983) as part of a
legitimation of land claims. So far, the park has not responded to the
idea of constructing a church there, but rather sees the potential to
develop the area as a "cultural heritage" site for tourists' consumption.

The state's denial of local customary practices is one aspect of how
park policies violate village moral economy. Alleged mistreatment at the
hands of the park guards, whom one farmer complained "come only to
harass us," compounds the loss of resource access.[25] There is a level of
fear and mistrust of the park guards, and a number of villagers made
serious accusations of abuse.[26] A villager said that after the right-of-way
was closed, the guards would beat people and rape the women that they
caught inside.[27] Others talked about rape that occurred if a woman was
caught alone collecting fuelwood, and complained in general about
beatings and threats with guns.[28] One farmer related that when guards
come to the village for an investigation, "[t]he guards talk to us like we
don't have any brains. They tell us, 'Sit and don't move.'"[29] One villager claims that he became the victim of harassment because he complained to the Usa River police station about park guards bringing a
poached buffalo into the village for sale.[30] He said he was in his plot on
the boundary, building a cage for his chickens, when the guards arrested
him for building a trap to catch scaly francolin (*Francolinus squamatus*)
inside the park. He was in the process of fighting the charges in court

when I talked with him. Similarly, the two park guards who were arrested for trying to sell buffalo in Ngurdoto Village (mentioned in the preceding chapter) were actually caught by the villagers. Curious as to why the villagers would attempt to arrest guards selling meat to a local butcher—a fairly common practice—I asked a research assistant to make local inquiries about the incident. He was told that these particular guards were rumored to be perpetrating a pattern of violence (including rape) directed toward villagers. Finding them bringing poached meat to the village merely gave villagers the opportunity to rid themselves of these "bad apples."

The villagers also complain that the guards are too often overzealous in carrying out their duties, leaving little room for mistakes. A farmer argued that most fines were unreasonable: "It is not good since the cattle don't enter the park purposely to graze, only because there is plenty of tall grass, so they cross just accidentally. Now, we don't feel that it is right."[31] One Ngongongare resident described why he thought that the park's policies are unfair: "If an elephant comes into my *shamba* and pulls up my banana plants, the guards don't do anything." He lifted up one foot and pointed his toe over an imaginary line: "But if a cow takes one step inside the boundary, the guards come and fine the owner."[32] In short, too often the guards were inflexible and left no space for negotiation. As will be seen below, the claim that cattle are "lost" or have gone into the park "accidentally" is the most common explanation for livestock trespass, and one that sometimes allows villagers room to bargain over the severity of the resultant fines.

To return to the problem of crop-raiding, this phenomenon is perhaps the most strongly felt violation of local moral economy. Predation by park animals puts a tremendous strain on household subsistence. Here is a small sample of replies to my questions about the amount of crops destroyed:

All of the farms nearby. The yearly production in each family's farm is destroyed by wild animals.

Almost the whole farm each year.

Many times I plant seven acres but harvest maybe one and a half to two acres.

They can destroy every acre.[33]

These accounts may sound exaggerated, but I observed numerous farms that had been virtually completely trampled and eaten. Often farmers would see me walking through the village and pull me aside to tell me agitatedly about the destruction done the previous night. Everyone who farms in the vicinity of the boundary suffers from predation by the park wildlife (see figure 10).

Yet villagers perceive that the park does little or nothing to alleviate the problem. Game scouts, villagers say, have too large an area: "They come maybe three days after the complaint and often they do not have bullets or even a gun."[34] Some say game scouts will not help in any case, unless they are offered a bribe. Not all villagers, however, make such claims, and there appears to be some recourse to appeal to authorities: "They help if you report to the Game Division. But from the park it depends on the warden. He can allow rangers to come to clear them or refuse, saying you should ask for the game scout."[35]

Nonetheless, there is a generally perceived lack of enthusiasm and capability on the part of the game scouts. This, combined with the park's "no-shoot" policy, has led villagers to accuse the government of caring more about wildlife than people. People commonly expressed the belief that the park administration, much as the colonial administration before it, is trying to drive people off their lands by allowing crop-raiding wildlife to go unchecked. Commenting on the lack of assistance from the park, a farmer told me, "They don't come to help us with this problem of animals, so why should we help them? We aren't going to help them catch poachers."[36] Ultimately, both people and wildlife suffer, because the local residents who bear the social and economic costs of the park have little enthusiasm for supporting the protection of the very animals who threaten their subsistence.

Villagers sometimes recall the "good old days" when the park was more responsive to their plight:

In 1977–78 they were helping by putting rangers out to guard the farms, and [they] erected a fence, but now, nothing. I don't know why they don't help, maybe because the manager has changed.[37]

There was a time when they were helping with an electric fence, but for a long time now they have not helped us. The problem was the fence wasn't being repaired frequently, and there was no caretaker and the battery ran down. Afterward the animals just walked right through the wire.[38]

These memories are not to be dismissed as nostalgic fantasies. In general, the park administration in the early years was much more accommodating to community subsistence needs than it was in the 1980s. A former park guard remembered, "A long time ago I used to allow them

Figure 10. The cultivation plots of Seneto Village abut the park boundary in the left foreground. (R. P. Neumann.)

[to collect fuelwood]. I used to allow people twice a month only; just the logs lying in the forest rotting. This was stopped in 1968, after Vesey-FitzGerald left. I just got a letter in 1968 from the park administration not to allow any person to enter the park to collect fuelwood."[39]

As strongly as villagers assert that the government does not answer to the problems caused by the park, the park authorities accuse the villagers of not upholding the law. Government officials view the villagers' lack of enthusiasm for park policies as a demonstration of their ignorance of the value of wildlife conservation. Much like their colonial predecessors, state authorities present an implied and often explicit image of villagers as either backward peasants or as criminals. Officials blame the villagers for the decline in wildlife populations, either directly as poachers or indirectly for failing to report the activities of poachers coming from outside the area.

Peasant farmers living near the park have formulated their own discourse, elements of which are expressed above, which contests that of the state. When talking of the park's policies, villagers have strongly held and readily expressed counter-notions of legality and justice. For example, a former ten-cell leader I spoke with was adamant that the park had no legal authority to prohibit access to the right-of-way through the park: "If the closure is legal, why was there no notice from the government? Why didn't they post any closure signs where the path enters the park?"[40]

Residents often claimed that it was the park guards and game scouts who were responsible for the loss of commercially valuable wildlife. They pointed out that there had been an abundance of rhino in the area prior to the establishment of the park. They countered that it was the soldiers and police hired as rangers from outside the area who brought knowledge of how to profit from commercial poaching. As one villager observed dryly, "The rangers are the ones with the guns."[41] An elder from Nasula explained, "Before the park started we used to graze our cattle with rhino. Then when the park was established, instead of taking care of rhinos, the rangers were the first ones to kill them. The park staff were involved in the loss of the rhinos and not the villagers. Even the park wardens who were transferred to this area were involved in poaching."[42] Another villager claimed that the rangers do most of the poaching in cooperation with outsiders (*wageni*). He had watched them come and go from his plot on the boundary and claimed to know their routes and methods.[43] Finally, the village chairman of Seneto, arguing the irrationality (in terms of wildlife protection) of the right-of-way closure, declared that there were rhino before it was closed and that they disappeared afterward. "If the Wameru were responsible for killing the rhino, there wouldn't have been any for the park to protect in the first place. The animals started disappearing after the park."[44] Regardless of who is to "blame" for the disappearance of rhino, there is an inherent truth to these statements. Ironically, the park lands were much more intensively used four decades ago, both by European settlers and surrounding Meru peasant communities, than they are now. Yet at that time, the area boasted one of the highest population densities of rhino in East Africa.

Embedded in villagers' counter-discourse is an historically rooted mixture of bitterness and belligerence toward the state and its conservation efforts. To elaborate, much of Meru peasant understanding of the conflicts between themselves and the park is derived from their historical experiences with the colonial state. Throughout Meru, it was not uncommon for my conversations with villagers to turn spontaneously to the topic of the forced eviction from Ngare Nanyuki. The eviction and the way Meru leaders organized a response of passive resistance were critical in shaping local political consciousness. Several people offered eyewitness accounts of the burning of people's houses. Others talked about the abuses suffered at the hands of Boer settlers and how, at the height of the disturbance, they were not even allowed to walk through the settler area. One such retelling was particularly striking to me because of the context in which it occurred. I was talking with an elder in Nasula about the park's rumored expansion when she explained to me: "This is our home and we are not going to be moved again." At this point she began to talk defiantly of their eviction from Ngare Nanyuki almost forty years earlier. "The leader of the white settlers put a gate down by the edge of the forest and another gate on the other side of Ngare Nanyuki. No Meru were allowed to pass even to go and see their families. But this man died, and we are still here. Now the park wants us to move again, but they will go before we do."[45]

Clearly, if villagers are drawing parallels between the national park and colonial repression, it raises questions about the effectiveness of wildlife conservation policies and underscores the severity of the conflict. As a local teacher whose family farm was partly taken over by the park expressed bitterly, "Do you think we have independence [*uhuru*]? Isn't this like colonialism [*kama ukoloni*]?"[46] The humiliation and deprivation that people living on the park boundary experience cannot do other than resurrect memories of the worst injustices of the colonial government. It is also of major symbolic importance that the park lands where villages directly abut the boundary were formerly settler estates. Rather than seeing these lands opened for Meru grazing, settlement, and other uses, after independence villagers watched the government make the former estates even more completely off-limits.

The parallels between the park and colonialism drawn within local discourse bring us back to the story with which I opened this chapter. Resistance to the loss of access continues to operate in Meru communities much as it did in response to the land alienations and state natural resource laws during the colonial era. Resistance continues, I have argued, because national park and wildlife conservation policies violate village moral economy. Policies and practices produce violations because they deny customary claims of access and threaten villagers' right to subsistence. Furthermore, village moral economy is breached within the context of complex and sometimes paradoxical day-to-day relations between park staff and local community members. It is the day-to-day relations that I focus on in the following section, exploring both the implications for community access to resources and the way the terms of a local moral economy are constructed through social practice.

The "Moral Community": Tensions over Inclusion

This section examines the dynamics of the relationship between the park staff and villagers, and explores its contradictions. The complex social relations that develop between conservation officials and the communities in which they work sometimes make it difficult for them to carry out their duties. The state's approach to wildlife conservation on the ground is partially concerned with severing relationships between park officials and villagers, or at least reducing the possibilities for such relationships to develop. While state officials attempt to confine the relationship to a bureaucratic, transactional nature, villagers maneuver to frame it as a "moral" one (see Bailey 1987). More precisely, villagers attempt to build social relations of reciprocity with park officials which are judged by the normative standards of the local moral economy. In a sense, this constitutes another layer of resistance to state modes of resource control. Ultimately, as will become clear below, the result for Meru villagers is a situation in which access to land and resources is unstable and unreliable.

Notions of landed moral economy, and characterizations of peasant society more generally, tend to be constructed upon the problematic assumption of a socially and geographically bounded community (see, e.g., Wolf 1966; Hyden 1980; Bailey 1987). For example, Bailey's concept of a "moral community" whereby "insiders" are included in a set of moral relationships and obligations while "outsiders" are excluded implies a questionable idea of isolated, self-contained peasant societies. Questions immediately arise. How are boundaries formed and maintained? Can individuals circulate between outsider and insider status and if so, how? Is this a model of social exclusion or inclusion? To elaborate, Bailey states that state officials are "unambiguously beyond the moral community of the peasant" (285). The category of state officials includes, as Bailey recognizes, a range of government employees, all automatically and "unambiguously" considered to be "outsiders." With this conceptualization, we are forced to accept the notion, now being increasingly challenged (Newbury 1994), that civil society is separate from the state. In actuality, government officials often have family, lineage, professional, or class ties within the locale of their operations. Ambiguity, in such cases, is inherent in relations between officials and citizens.

A more amorphous and dynamic (and therefore more reflective of social practice) conception of the "moral community" is one that encompasses "the entire peasant universe in ever widening orbits of responsibility from the household, to extended kin, to village patrons, and ultimately to the state itself" (Watts 1983, 106). At each level of encounter, from face-to-face negotiations to anonymous bureaucratic transactions, different tensions will arise. These different tensions are to a large degree defined by the nature of class relations. To borrow from Watts's analysis of Hausa moral economy, reciprocal obligations operate in two dimensions: horizontally among equals and vertically among unequals, such as between officeholders and commoners (123). The type of obligation, the manner in which it is formulated discursively, and the possibilities for its enforcement vary greatly between horizontal and vertical relationships. Finally, the tensions in Meru villages near the park, I argue, are not over who is *excluded* from the "moral community" but over the reluctance of state officials to be *included* in relations of mutual obligation.

Within the social and political context of the surrounding villages, the Arusha National Park administration functions somewhat as a large estate owner, one whose claims to the land lack legitimacy. It is important to note here that landlord-tenant relationships did not exist in precolonial Meru. Under colonialism, however, squatting on settler land became common in the area, either illegally, as an assertion of customary claims, or legally, as an economic strategy. The chapter's opening story indicates that squatters and laborers had an expectation of paternalistic behavior on the part of these estate owners. In short, the story represents an attempt by the weak to engage the wealthy in a system of patronage based on their notion of justice and morality (Scott 1976; Hyden 1980; Bailey 1987).

I suggest that the formal appeals (letters, delegations, and meetings) from villagers to government authorities for access to the park's land and resources function as one means to articulate local notions of rights and obligations. In this they are quite similar to the sorts of appeals to settlers mentioned in the story with which the chapter began. As we have seen, however, the park administration does not recognize a relationship with villagers based on ideas of mutual obligation or paternalism. Park wardens have consistently refused to grant access to critical grazing resources during times of dearth. Yet, within the boundaries there are grazing areas, wood supplies, forests for honey production, and water sources that were once important elements in the Meru system of production, and which were taken over by the state without compensation. Most villagers suggested that the park was of little direct benefit to their community, and that the tall grass and dead wood inside were going to waste. Some argued that villagers should be allowed in to take advantage of the wealth of the park: "It could be helpful to have an area to graze our herds, like [the Maasai are allowed to] in Ngorongoro. There are very good areas for grazing, such as Momella and Trappe Valleys. Also the firewood would be a great help."[47] Park authorities remain deaf to these appeals. Efforts to establish a relationship of reciprocity, to envelop the park bureaucracy in the local moral economy through official petitions, have not been successful. Land and resources inside the park were once accessible through locally constituted social relations, but are now formally controlled by a bureaucratic state.

Clearly, the administration wants to restrict its involvement with villagers to a formalized, unidimensional relationship, citing the national legal code in refusing any obligation to assistance. The relationship between neighboring peasant farmers and the park authorities is reminiscent of Bailey's generalized argument that "it is the supplicant who seeks to make the relationship diffuse: to make it a moral relationship . . . [I]f the official or politician is dominant he will try to retain the transactional character of the relationship and not have the sharp edge of the bargain blunted by moral considerations" (1987, 286). The state, however, is intent on bureaucratizing national parks administration, restricting staff loyalty to impersonal and functional purposes.

The unofficial relationships between the park staff and the villagers are of a different character. The daily lives of staff members and villagers are intertwined in complex and sometimes contradictory ways (see figure 11). In many ways the villagers and staff cooperate and coexist on a daily basis. The villages near the headquarters have free use of the dispensary in the park, and the families of the park staff go to the village churches and send their children to the same school that the village children attend. In the evenings, the young men of both groups form a local soccer team that competes against teams from other villages. The rangers and wardens buy their provisions in the local markets and have their maize milled in the villages.

On another level their relationship is antagonistic in the extreme,

Figure 11. Park guards clown with a Nasula resident, playing "catch the poacher" for the author's camera. (R. P. Neumann.)

considering the stories of beatings, humiliation, and harassment at the hands of the park guards. Villagers complain that the park management is not concerned with the difficulties they face living next to the park and that the wardens only come to the village when they are enforcing the laws. Summing up the antagonistic aspects of their relationship, a villager told me, "The guards are our enemy and to them we are the poachers. If I cross into the park they will shoot me. Are they not my enemy?"[48]

Despite these hostilities, the rangers are at times dependent on the goodwill of the villagers for their livelihood. Life is difficult for wage earners anywhere in Tanzania, but park staff have unique problems. At the time of this study in 1990, park guards were earning thirty-six hundred Tanzania shillings per month to start, equivalent to about $18. Home gardening and the keeping of livestock, practices that supplement many Tanzanian workers' incomes, are prohibited inside the park, so employees must use their salaries to purchase nearly all their food requirements from villagers. Thus there is a great temptation to trade favors—like looking the other way when cattle trespass—for food.

From the mid-1970s to the early 1980s the temptations were probably greatest because park funds were chronically in short supply and at times salaries could not even be paid.[49] Villagers told me that toward the end of the fiscal year when the budget had been exhausted, rangers would come to the village seeking food. A resident described the situation to me by relating a mock exchange between himself and a ranger: "If one of the rangers comes to the village and says 'Mzee, I'm hungry and have no food,' I say, 'Why did you refuse to let us get fuelwood?' And he will say, 'I think you can come to the park anytime.'"

The park guards are therefore susceptible to the villagers' efforts to engage them in a moral relationship of mutual obligation and reciprocity. For villagers who have bargaining power, often in the form of surplus crops, there is a distinct possibility of making "arrangements" with the rangers for access to park resources. A resident living near the boundary explained to me that "[t]here is plenty of grass in the forest; you can negotiate with the rangers to get permission to send your cattle inside. It is not allowed, but you can negotiate, because you are living near the forest."[50] The details of these bargains are, for obvious reasons, closely guarded secrets, even among neighbors. Usually the bargain involves an exchange of staple foods or milk for the chance to collect fuelwood or to graze cattle. Sometimes they are relatively public. In Nasula Village near the park headquarters, every Saturday a park truck would deliver a load of firewood from the park to a woman's house in exchange for the warden being supplied daily with fresh milk. Even if people are caught in violation of park laws, there is a chance to negotiate to minimize the cost. Many villagers reported to me that they had been fined for having livestock in the park but managed to bargain directly with the rangers without having to go through a formal legal process. Two farmers in Ngongongare said they often had to pay fines, mostly for cattle, and that these fines were not fixed.[51] That is, if they are caught on the spot, they can haggle with the ranger and pay without a receipt. If the cattle have to be kept overnight, they usually then go through the official process of paying the fine. Many people thus talk of their children "losing" their cattle in the park "accidentally" as a way of opening a space for negotiations with the guards (see figure 12).

The development of a moral relationship between park staff and villagers, in short, inhibits the effective implementation of state wildlife conservation policies. The government currently utilizes a number of strategies to reduce the chances for villagers to build this type of relationship. First, it follows a policy of frequent inter- and intrapark

Figure 12. Two young girls herd their family's cattle along the boundary of the park (background) and Nasula Kitongoji. (R. P. Neumann.)

rotation of personnel, reducing the possibility for ties of friendship to develop. Second, park guards are not hired from the local population. Personnel are brought from outside the area to circumvent the problem of trying to sever ethnic or familial loyalties. As an example of what the government seeks to eliminate, one Meru farmer who worked as a guard in Arusha National Park in the early years told me that people in the villages "loved" him because he used to allow them to bring their livestock into the park.[52] Finally, when hiring park rangers, the government does not necessarily look for strong conservation ideology, but rather selects people trained in the use of violence. Guards and rangers are recruited from the National Defense Force, the police, and prison guards.

Even though the social lives of the villagers and park staff are intertwined in numerous ways, and the guards are at times in a position of dependency in relation to the villagers, the villagers have not been successful in engaging them completely in a reciprocal relationship. This is due partly to the guards' dualistic social life as both community members and enforcement agents for the state. Just as important, however, are the government's efforts to limit the possibilities for a relationship to develop, with one consequence being that park guards act inconsistently, sometimes operating according to the obligations of a "moral" relationship, at other times according to the formal and impersonal rules of modern bureaucracy. The result for the villagers is summed up clearly by a farmer who explained to me, "The rangers depend on the villagers for their livelihood. They come to us for milk, beans, and maize. But if there is any trouble with the park, they act like they don't know us."[53] The tension between the groups here is not over the exclusion of "outsiders" from the local "moral community," but over the outsiders' refusal to become fully integrated community members.

Within these shifting relationships of cooperation and animosity, dependence and autonomy, the Meru villagers make informal and illegal arrangements to gain access to resources. But these arrangements are risky and unreliable, because there are penalties if one is caught, and because villagers cannot depend upon the guards to hold up their end of the bargain. The new forms of land and resource allocation and control that emerged when the state entered the picture are unstable and unpredictable. Wardens and rangers are shifted frequently between posts. Sometimes villagers and staff are enemies, sometimes friends. Some villagers have the means to bargain for access, others do not. Unlike customary controls over access, there are no reliable rules or patterns of social behavior that would allow a household to plan for and depend upon the use of resources inside the park. In addition to the elements of instability and unpredictability, the moral economy is in a sense inverted. Where the moral economy operates to assure a minimum level of subsistence to even the poorest members of the community by providing access to communal lands and resources, in this case only those well off enough to have something to bargain with can gain access to park land and resources.

The nature of the relations between the park and surrounding villages does not bode well for wildlife. Although there is no ecological monitoring program within the park and almost no data exists on fluctuations in wildlife populations, some conclusions can be drawn concerning the state of wildlife conservation. The clearest conclusion is that the park has basically failed to protect populations of large mammals. This is especially true for commercially valuable species, such as elephants, whose numbers have fallen dramatically, and black rhinoceros, which have been extirpated. Though commercial poaching for horn and tusks was apparently conducted mainly by people from outside the area, it would be difficult for it to take place without the knowledge of local villagers, if not their complicity. In fact several villagers made clear to me that they knew who the poachers were and how they operated. Since villagers find that the park is of no benefit to the village, and is often harmful, they have little interest in cooperating with state agencies in wildlife conservation efforts. Despite conservation authorities' claims to the contrary, residents believe that the government places the rights of animals above those of humans. Policies would seem to support this interpretation. The ramifications for community support are illustrated by a villager who told me: "If I heard a gunshot, or even if I saw someone shoot an animal right in front of me, I wouldn't tell them anything. The national park doesn't care about wild animals eating my crops; why should I care about their problems?"[54]

Colonialism and the accompanying land alienations for European estates has interpenetrated another wave of meaning into the landscape of the park that deeply influences the Meru's interpretation of the struggle today. For the Meru living near the park and dependent upon its resources, the seizure of land by the independent state for conservation differs little in practical and symbolic terms from the initial loss of the same land to European estates. The Western bourgeois sensibility of "nature appreciation" is not a universal value, and the ideals of "Eden" supposedly embodied by Arusha National Park are, as yet, not a part of local consciousness in Meru.

For many Meru peasants, the loss of access to the land and resources inside the park is part of the larger historical process in Tanzania of the replacement of local, customary authority by centralized, state authority. The implementation of state wildlife conservation laws introduced not only a new set of social structures and institutions for controlling access to natural resources, it initiated a new dynamic within social relations that works against the interests of Meru peasants. It also imposed a new set of meanings on the land, a landscape of nature consumption, devoid of human history, that clashed with locally constructed meanings.

From the perspective of state officials, the villagers' activities are cause for alarm since they threaten park management goals. For park administrators, the root causes of these "threats" to the park include criminal intent, population growth, and a lack of understanding about conservation by local residents. Administrators also see themselves lacking sufficient management capacity for responding to the challenges to authority. A recent Tanzania National Parks report cautions that there is also "an even more relentless threat, and that is the growth in the number of people inhabiting villages on the periphery of the national parks."

One solution is "to educate the masses of people in surrounding villages, to teach them that wildlife has an important part to play in the national heritage."[55] If park management were able to solidify its position by, for example, clearly marking its boundaries, then "[t]he present conflicts, caused when people enter the park for grazing or other illegal activities on the pretext of either not knowing the boundary or denying the passage of boundary lines because there are no clear markings, will be alleviated."[56]

If the above passages imply that authorities view local residents with a mixture of suspicion and frustration, then the reports from the field confirm these sentiments. As revealed by the park reports, one need only step over the boundary line to be transformed into a "poacher": "Another poacher was arrested at the same area for entering the park without permission."[57] Lack of sympathy for the park's management problems is interpreted as a sign of local guilt: "This silence by the village leaders has led us to think that they are cooperating with poachers."[58] According to officials, the villagers' unwillingness to cooperate boils down to the fact that "few people are aware of the national park's importance."[59] The obvious solution is to help them see the light. "At this time action taken to reduce poaching was to visit and educate the ten cell leaders on the importance of conserving the environment by preventing livestock from getting in the park and destroying it."[60] What conservation officials recognize as a lack of awareness and education is more profitably interpreted as a defense of customary rights of access and a struggle over the meaning and representation of the mountain's landscape.

The park administration sees Meru peasants "encroaching" on the national park. By viewing the situation from the "bottom up" (Blaikie 1985; Blaikie and Brookfield 1987), however, and within the historical context of state-mandated changes in land and resource use, it is the park that is seen to be encroaching on the Meru villages. For Meru peasants living near the park, the expanding park boundaries and legislated restrictions on movement and access to essential resources are the moral equivalent of colonial land alienations. After nearly one hundred years of state-directed forest and wildlife protection on Mount Meru, there is little popular support for conservation laws.

The subculture of resistance to conservation laws that so frustrates the park's enforcement efforts is misunderstood by authorities as an ignorance of the value of wildlife protection. Even a cursory analysis of local discourse reveals that opposition is aimed not at conservation per se, but at the way the policies are designed and implemented, which for the most part is contrary to villagers' interests. For example, a young farmer born in Nasula declared, "The park is a benefit for us because our youth can see the game animals without having to travel. Arusha National Park also provides the government with an income." The direct impact of the park on her life, however, creates ambivalence: "[O]n the other hand, the park makes life worse because of the wild animals destroying our crops."[61] People are well aware of the purposes of the park, as another villager explains: "We know that the park keeps the wild animals for future generations and it earns a lot of foreign currency for our government, but we are asking the park management to discuss with us how they can help us on this problem of wild animals destroying or eating our crops."[62]

Villagers time and again expressed to me the superiority of local knowledge of natural history, ecology, and local practices over that of the park managers brought from the outside. Several people stressed that the park has to start working with the villagers because it is not the warden or the rangers that know the park, but the Meru, whose land it is. They know all the poachers in the village and know the trails they use: "We know who the poachers are. I know a couple of them myself. Most of the poachers are just trying to keep the animals out of their farm plots and a few are doing it to sell meat in the village."[63] The park rangers, however, are less familiar with the local physical and social terrain. One resident explained the difficulty that staff from the outside have at Arusha National Park: "Serengeti is very open, but here there is thick bush and thick forest. When they come here it is like they are in the dark."[64]

Since independence, the Meru have continued to lose ground in terms of control over critical resources. Peasant farmers on Mount Meru are largely shut out of the formal political process and are left to whatever defense they can muster, de facto alliances with poachers, foot dragging in regards to compliance, and piecemeal tactics to strengthen customary claims to land and resources. By either social or ecological standards, wildlife conservation and nature protection on Mount Meru are failing. The conflict at Arusha calls into question the sustainability of a rigidly defined national park model in Africa, where human activities are absolutely prohibited and where control is placed entirely in the hands of the state. Solutions to these types of conflicts must address the variability in deeply held symbolic meanings attached to land, as well as questions of material access. Not only can this be done while still maintaining the protection of wildlife and habitats, but without attention to these questions, the efforts of Tanzanian and international conservationists are unlikely to succeed.

Epilogue

When I began this study in 1988, various wildlife conservation interests across Africa were becoming increasingly occupied with addressing the types of conflicts that were plaguing Arusha National Park. Up to that time, there had been sporadic, isolated attempts to try to redress the loss of access to livelihood resources that resulted from the establishment of national parks (see, e.g., Western 1982; Lindsey 1987). By the first half of the 1990s, these efforts had been institutionalized within African countries' natural resource agencies and within international conservation nongovernmental organizations (NGOs). A plethora of new acronyms sprouted on the conservation landscape around the turn of the decade: ADMADE (Administrative Management Design for Game Management Areas) in Zambia; NRMP (Natural Resources Management Project) in Botswana; CAMPFIRE (Communal Areas Management Programme for Indigenous Resources) in Zimbabwe; and CCS (Community Conservation Service) in Tanzania.

These programs, along with an assortment of other nonprogrammatic initiatives, seek to promote "local participation" and "benefit sharing" as part of a comprehensive strategy for biodiversity protection. They are known by a variety of labels, including community-based natural resource management (CBNRM), integrated conservation development projects (ICDPs) and, community conservation (CC), protected area buffer zones, or buffering strategies (Bergin 1995; Bloch 1993; Neumann 1997a; Omo-Fadaka 1992; J. Sayer 1991; Wells and Brandon 1992; 1993). Most are founded on the idea that conservation and development are mutually interdependent and must be linked in conservation planning (see Kiss 1990; McNeely and Miller 1984; Miller 1984). The underlying reasoning is that neighboring communities must receive direct benefits from protected areas in order for conservation policies to be effective (Gibson and Marks 1995; Barrett and Arcese 1995). Benefits to local communities include those directly related to wildlife management (wages, income, meat), social services and infrastructure (clinics, schools, roads), and political empowerment through institutional development and legal strengthening of local land tenure (Ghai 1992; Gibson and Marks 1995; Makombe 1993).

In recent years one of these programs, Zimbabwe's CAMPFIRE, has become an icon among international NGOs and government conservation agencies. The roots of CAMPFIRE can be traced to a shift in Zimbabwe in the early 1960s from strict game preservation to sustainable game ranching, later formalized in 1975 in the Parks and Wildlife Act (Murphree 1997). The purpose of the legislation was to confer the ownership and management of wildlife on the (white) owners of alienated lands. Then in 1978 the Department of National Parks and Wild Life Management initiated WINDFALL (Wildlife Industries New Development for All), a program designed to give meat culls from parks and revenues from safari hunting to neighboring district councils in Zimbabwe's communal areas. Its central focus was therefore the economic utilization of game, rather than wildlife conservation or protected area management per se (Murphree 1997). Though WINDFALL failed, it laid the foundation for CAMPFIRE after a 1982 amendment to the Parks and Wildlife Act extended economic benefits from wildlife to the communal lands. After long negotiations within and between government branches (Murphree 1997), CAMPFIRE was officially launched in 1986. The agreement provided for de facto devolution of management and revenue of wildlife resources to production units, naming the Regional Development Councils as "appropriate authorities" for receiving and distributing benefits.

Though CAMPFIRE is ostensibly about wildlife utilization and the redistribution of revenue, it does have relevance for protected area management. In some cases, it appears to be part of an effort to mitigate past conservation displacements. Several of the CAMPFIRE projects involve communities on national park boundaries that were displaced in the creation of protected areas or by dam projects (Bird and Metcalfe 1995, 3; Murphree 1995, 9; Taylor 1995). In other cases, CAMPFIRE appears to be part of a buffering strategy. Child notes that WINDFALL, the precursor to CAMPFIRE, was originally motivated by the NPWLM Department's fear that the protected areas in Sebungwe region "would become ecological islands threatened by settlement" (1996, 361). The western portion of Tsholotsho District—where the district planned its wildlife area as part of CAMPFIRE—shares wildlife populations with bordering Hwange National Park (Thomas 1995). Thus CAMPFIRE appeared to be a good model for addressing conflicts between protected area management and neighboring communities.

Whatever the explicit intent or subsequent effects, CAMPFIRE has spread rapidly in Zimbabwe from nine participating districts in 1989 to twenty-five in 1995 (Murphree 1997). Partly because of the potential for multiple readings of CAMPFIRE's agenda—from game ranching to protected area buffering strategy to CBNRM—the idea has also caught fire across the continent. Through a string of workshops, conferences, and consultations, often sponsored by major international conservation NGOs and donor agencies, the philosophy of CAMPFIRE—if not the precise policy mechanics and tenure arrangements—has been embraced in other African countries (see, e.g., Wilson 1997). Included among these is Tanzania's CCS program.

Though CCS is not a direct descendent of CAMPFIRE, it can be viewed as part of the widespread programmatic shift in conservation that CAMPFIRE has come to symbolize. While it turns the focus of protected management toward rural community development, CCS, which is administered within TANAPA, is fundamentally different from the more widely known CAMPFIRE. Unlike CAMPFIRE, the impetus for CCS came from outside the country in the form of financial and technical support from the African Wildlife Foundation (AWF). Also, CCS does not involve the devolution of tenure, de facto or otherwise. Patrick Bergin (1995) categorizes the CCS program as a community conservation (CC) effort in contradistinction to CBNRM because it does not address the issue of resource management on community-controlled land. Nor does it permit access to or utilization of park lands and natural resources. The heart of the program is the return of a share of park tourism revenue directly to the communities surrounding the national parks.

CCS has important implications for the historic conflicts between state resource management agencies and the Meru communities sur-rounding Arusha National Park. After an initial pilot program near Serengeti National Park from 1985 to 1990, CCS was expanded in 1991 to include Tarangire, Lake Manyara, and Arusha National Parks.

Though Arusha was of little interest to wildlife conservationists, TANAPA officials wanted it included in the program expansion. Bergin noted that Arusha "came to be thought of [by TANAPA officials] as a 'problem' park where CC activities could possibly reduce pressures" (1995, 28). Arusha was thus one of the first parks to receive a full-time CC warden in 1991. Also at this time, the TANAPA Board of Trustees created a budget fund that allowed the agency to direct tourism revenue to small-scale development projects in the neighboring communities of the four parks.

In 1992, Bergin, who had been employed by AWF as an advisor to the CCS program, conducted doctoral dissertation research in the communities adjoining Arusha National Park in order to assess the effects of CC (Bergin 1995). His findings indicate a significant shift in the park administration's approach to neighboring communities and a renewed attention to the conditions of underdevelopment resulting from national park policy. As part of an effort to establish dialogue, the park stationed rangers to help guard crops at night, whom villagers housed, fed, and assisted. The park used its first allocation of special community funds—called Support for Community Initiated Projects (SCIP)—to build a classroom and a teachers' office in Ngongongare. A rise in tourism at Arusha National Park gave a boost to the CCS program at a crucial time. Bergin noted that noncitizen visits grew from 2,365 in 1987–88 to 11,301 in 1993–94 (1995, 29). This provided the foundation for expanding local opportunities for wage labor as porters for tourists climbing the mountain. The locally organized porters union had eighty-six names in 1994 (Bergin 1995, 128). Porters earned an average of Tsh 3,000 per day ($6 in 1994) on a journey typically lasting three days (Bergin 1995, 131). In combination with tips averaging about Tsh 2,500 per trip, porterage has become a crucial source of supplementary income in a country where the minimum wage for private employees was only Tsh 5,000 per month in 1994. Most of the porters come from Olkangwado Kijiji, where income sources are extremely limited.

Bergin's positive assessment notwithstanding, empirically measuring the "success" of programs like Tanzania's CCS is a difficult methodological and conceptual problem. The efforts at dialogue that the CCS program initiated are undoubtedly important for reducing the level of animosity between the park administration and surrounding Meru communities. Whether or not this has a positive effect on protected area management is not so clear. What, for example, is the cause-and-effect linkage, if any, between the building of a classroom and the reduction of grazing trespass? It is also difficult to separate the results of CCS initiatives from events and processes external to the program. The growth of opportunities for wages from porterage, for example, though supported by CCS, is mostly a result of the overall growth in Tanzania's share of the regional tourism market. Finally, it is clear that the program has severe limitations for addressing the demands for recognition of customary claims on the park's land and resources documented in preceding chapters. Access to park lands and resources is explicitly not part of the program. In no situation is this stance more clear than in the case of the legislatively designated three-meter-wide right-of-way through the park. Though specified in the Parliamentary acts that created the park, Bergin reported that community requests to reopen the path were once again denied by the director general of TANAPA (1995, 76).

Two conclusions can, however, be drawn from this early assessment of CC at Arusha National Park, one discouraging, one hopeful. On the discouraging side, CCS does not offer even the possibility of recognizing customary claims within national park boundaries. There is no suggestion of any retreat from the spatial segregation of nature and society nor of any sort of comanagement. The park will continue to function as a landscape of consumption for the aesthetic pleasures of tourists, principally from Europe and North America. The onus is on the Meru peasant communities to abandon the locally constituted meanings inscribed on the mountain slope, to forget the ancestral sites and the meeting places. On the hopeful side, the inclusion of Arusha National Park in the initial CCS program signals the potential power of local protest and everyday acts of resistance to influence state policy toward recognizing and redressing the costs of state-led nature preservation borne by the peasantry. Arusha was viewed as a "problem" park precisely because of the decades of resistance to the loss of access resulting from reserve and park establishment. As detailed in chapter 6, one of the main avenues for Meru communities trying to regain lost rights was by appeals and petitions through official channels. Bergin noted that these "strategically channeled complaints are taken seriously by TANAPA" (1995, 28), as the director general is a presidential appointee who must stand before Parliament to defend the agency's budget. These sorts of tactics may become increasingly effective as "democratization" in Tanzania provides a climate of political accountability and new openings for grassroots mobilization among communities displaced by conservation (Neumann 1995a). Ultimately, the success and advancement of CC initiatives at Arusha and in similar situations will depend on the continued assertion of customary claims and demands for justice emanating from below.

Notes

Introduction

1. National parks and reserves refer to the IUCN international standard categories II and IV. Category II is "National Park," defined as an area "under state control . . . set aside for conservation . . . in which killing, hunting and capture of animals and the destruction of plants . . . are prohibited." Category IV is "Managed Nature Reserve/Wildlife Sanctuary," including game reserve, which is defined as "set aside for the conservation . . . of wild animal life and . . . habitat . . . within which hunting . . . is prohibited except by . . . control of reserve authorities, where settlement and other human activities shall be controlled or prohibited" (IUCN 1991, xiv–xix).

2. The main objection to the use of the term "Bushman," even in quotes, is that it has racist and derogatory connotations. In choosing to use the term I am agreeing with the argument put forth by Gordon (1992). First, using an alternative such as *San* or *Kung* or *Ju/wasi* (the most common substitutes) simply replaces *Bushman* with another generalized label that masks a great deal of cultural and geographic variation. Second, and more important, the origins of the label *Bushman* can be found in the history of local resistance to European colonialism in Southern Africa. It was a sociopolitical category "into which all those who failed to conform or acquiesce were dumped" (1992, 6). Gordon uses the term as a way to "make social banditry respectable again" (ibid.) as a form of valiant resistance to colonial occupation.

3. TANAPA, *Report and Accounts of the Board of Trustees, 1972–75.*

Chapter One

1. *The Journal of American History* 76, no. 4 (1990) is a special issue devoted to environmental history which contains many of the essential theoretical debates.

214 NOTES TO CHAPTER 2

2. This is part of the central argument put forward by J. M. Blaut (1993) in *The Colonizer's Model of the World*. According to Blaut, European imperialists legitimated the appropriation of non-European lands through the development of a "myth of emptiness" (1993, 25). The essential element in the myth being that colonized territory was devoid of population or, if it was populated, it was a population that was nomadic with no territorial claims, or that had not developed the concepts of political sovereignty and economic property. Thus, the lands could be morally and legally taken over by the more advanced and industrious European states.

3. The recent work of Nancy Peluso (1992) on peasant resistance to state control of natural resources on Java offers a comprehensive history of the development of state-controlled scientific forestry that exemplifies a typical pattern in colonial territories.

4. The risk-reducing logic of creating affective ties parallels F. G. Bailey's argument that peasants will try to engage officials and other community outsiders in a "moral relationship" (1987, 286) for precisely this reason.

5. Many of the ideas of conflicting interests between moral economy or economy of affection and a more impersonal, bureaucratized market economy imply but do not acknowledge a great debt to Max Weber.

Chapter Two

1. *Arusha District Book,* Tanzania National Archives (TNA); and Hans Cory, "Tribal Structure of the Meru" (TS), n.d., the Cory Papers, University of Dar es Salaam Library, East Africana Collection.

2. Interview with Sori Kibata [pseud.], 23 August 1990. The "prophets" referred to were local shamans who were respected for their power to dream and predict or recognize danger to the Meru. They were important advisors in major decisions such as the acceptance of a group of outsiders into their society.

3. Interview with Sori Kibata [pseud.], 23 August 1990.

4. Interview with Mzee Pallanjo [pseud.], 5 June 1990. (*Mzee* is a respectful title for elders.)

5. Interview with Sori Kibata.

6. Interview with Mzee Kaaya [pseud.], 7 September 1990.

7. Interview with Sori Kibata.

8. Interviews with Yesuria Ukuta [pseud.], 24 January 1990, and Sori Kibata [pseud.], 19 August 1990.

9. The historical boundaries of the Meru land claims were collaborated in interviews with elders of the Mbise and Kaaya clans.

10. Interview with Sori Kibata.

11. District commissioner (DC), Masai/Monduli, "Report on the Movement of Masai from the Commission Lands," 13 December 1951, Public Record Office, London (PRO) CO 822/430.

12. Interview with Sori Kibata. Njeku is the name given to the Mbise and Nnko clans' shared ceremonial site in the Mount Meru Crater.

13. DC, Arusha, to provincial commissioner (PC), Northern Province, 16 September 1930, TNA 45/9, Accession No. 69.

14. Tanganyika Territory, "Report of the Arusha-Moshi Lands Commission," Dar es Salaam, 1947.

15. "Veterinary Officer's Report," 1939, *Arusha District Book*, TNA.

16. DC, Arusha, to PC, Northern Province, 16 September 1930, TNA 45/9, Accession No. 69.

17. B. J. Hartley, agriculture officer, "A Brief Note on the Meru People with Special Reference to Their Expansion Problem," TS, 1938, TNA 45/9 Accession No. 69.

18. "Arusha District Agricultural Officer's Report," 25 March 1938, *Arusha District Book*, TNA.

19. DC, Arusha, to PC, Northern Province, 16 September 1930, TNA 45/9, Accession No. 69.

20. Tanganyika Legislative Council, "The Meru Land Problem," white paper, 1952, TNA.

21. A letter from the Tanganyika Coffee Growers' Association, signed by Maj. S. E. du Toit and others, recommends that Ngare Nanyuki "should be a homogeneous European block" and that Trappe's "Momella Farms Nos. 40 and 41 and Farms 325, 326, 328, and 329 should be allocated for African use." Tanganyika Coffee Growers' Association, "On the Proposed Redistribution of Alienated and Tribal Lands on and around Kilimanjaro and Mt. Meru," memorandum, prepared for submission to the Arusha-Moshi Lands Commission— September 1946, Rhodes House (RH), MSS. Afr. s. 592. box 4.

22. PC, Northern Province, 1949, *Annual Report*, TNA. The PC is referring to the plan to forcefully relocate three thousand Meru as part of the colony's "economic development."

23. PC, Northern Province, 1949, *Annual Report*, TNA.

24. District officer (DO), Arusha, to H. Fosbrooke, 23 November 1953, RH MSS. Afr. s. 1790.

25. Rogers, Colonial Office, to Hutt, Tanganyika Territory (TT) Secretariat, 25 August 1951, PRO CO 822/430.

26. Interview with Mzee Abraham [pseud.], 20 January 1990.

27. Most likely the settler, Trappe, did not support the eviction, because his European neighbors had suggested that his estate, Momella Farm, be confiscated and used for the resettlement of Meru from Ngare Nanyuki. Tanganyika Coffee Growers' Association, "On the Proposed Redistribution of Alienated and Tribal Lands on and around Kilimanjaro and Mt. Meru."

28. Interview with Mzee Kaaya [pseud.], 9 December 1989.

29. Interview with Mzee Abraham.

30. Gov. Twining, TT, to Lyttelton, secretary of state for the colonies (SS), 5 February 1954, PRO CO 822/806.

31. "An Appreciation of the Kikuyu Situation in the Northern Province on 27th October, 1952," confidential memorandum, PRO CO 822/502.

32. Gov. Twining, TT, to Lyttelton, SS, 8 December 1952, PRO CO 822/431.

33. "An appreciation of the Kikuyu situation."

34. Unsigned notes in the file, "Mau Mau Repercussions," PRO CO 822/502.

35. Gov. Twining, TT, to Lyttelton, SS, 8 December 1952.

36. Gov. Twining, TT, to Lyttelton, SS, 5 February 1954.

37. Gov. Twining, TT, to Gov. of Kenya, Baring, 15 December 1953, PRO CO 822/502.

38. Ibid.

39. TS of sworn affidavits and reports on political disturbances in Meru, TNA 16813/1.

40. PC, Northern Province, 1951, *Annual Report*, TNA.

41. Interview with Mzee Abraham.

42. Gov. Twining to SS, 3 October 1952, PRO CO 822/431.

43. DC, Arusha, to PC, Northern Province, 12 March 1952, TNA LAN/15/Vol. III, Accession No. 472.

44. Secretary Hinds, Usa River Planters Assn. to DC, Arusha, 12 February 1953, TNA LAN/15/Vol. III, Accession No. 472.

45. Focsaner (settler) to DC, Arusha, 19 June 1953, TNA LAN/15/Vol. III, Accession No. 472.

46. Speech by Gov. Twining to Meru Baraza at Poli, 21 October 1954, TS, *Arusha District Book*, TNA.

47. See Julius Nyerere's foreword to Japhet and Seaton (1967).

48. Based on values calculated by Spear (1994) plus data from the 1988 national census (United Republic of Tanzania 1989).

49. Interview D. Kalanjo [pseud.], 9 December 1990.

50. Anonymous survey respondent, Nasula, 16 September 1990.

51. The following information is from a series of interviews in Nasula with Mzee Molel [pseud.] and Mzee Simera on 18, 19, and 22 January 1990.

52. The following information is from two interviews in Nasula with Mzee Joseph on 24 and 25 January 1990.

53. Among both the Arusha and Meru on Mount Meru, the pre-European pattern of settlement expansion typically involved an individual moving into an uninhabited area and demarcating the land that he and his wives intended to cultivate and graze. The area was usually large enough to accommodate the needs of the next generation. This practice continued even into the immediate postcolonial period, when previously unoccupied lands on Mount Meru were settled or when abandoned estates were claimed by Meru.

54. Anonymous survey respondent, Nasula, 14 April 1990.

55. Interview with Mzee Perimu [pseud.], 11 June 1990.

56. Interview with Mzee Palanjo [pseud.], 8 June 1990.

57. Interview with Mzee Perimu, 11 June 1990.

58. Anonymous survey respondent, Nasula, 15 June 1990.

Chapter Three

1. Portions of the history of the Tanganyika Forest Department described below are based partially on material in Neumann (1997b).

2. R. S. Troup, *Report on Forestry in Tanganyika Territory, 1935*, TS, TNA 23115.

3. Tanganyika Forest Department, 1923, *Annual Report*, TNA File 1733/A/13/L: 88.

4. Ibid.

5. "Extract from German Ordinances and Decrees of German East Africa," TS, TNA 2708.

6. German East Africa Game Ordinance of 1908/1911, TNA 2708.

7. Tanganyika Forest Department, *Annual Report*.

8. Leechman to chief secretary (CS), 17 May 1920, TNA 2708.

9. Tanganyika Forest Department, *Annual Report*.

10. Tanganyika Territory, Forest Rules of 1921, TNA.

11. Notes on Draft Game Ordinance by C. F. M. Swynnerton, 1921, TNA 3260.

12. Swynnerton, game warden (GW), to CS, 7 August 1923, TNA 7227; Teare, GW, to CS, 22 April 1933, TNA 12005.

13. CS to PC, Mwanza, 21 May 1929, TNA Secretariat File 13371.

14. From the full title of Tanganyika Territory, 1940 Game Ordinance, TNA.

15. A full discussion of the establishment of national parks and the extent of the SPFE's influence on Tanganyika's national parks and wildlife policy will be presented in the next chapter.

16. Tanganyika Territory, 1940 Game Ordinance.

17. Governor's comments on Executive Council Circular, 2 February 1926, TNA 3260/2.

18. Extract from the summary of the Combined Agricultural, Cotton Entomological and Mycological Conference 1926, quoted in the *Proceedings of the 2nd African Game Conference*, Mombasa, 11–15 January 1927, p. 6. PRO CO 822/5/7.

19. Chief native commissioner, East Africa, to SS, n.d., quoted in the *Proceedings of the 2nd African Game Conference*.

20. Forest Department Circular No. 1 of 1933, 3 January 1933, TNA 21559.

21. Comments on Proposed Forest Rules of 1928, initials illegible, 13 August 1929, TNA 10580.

22. CS to PC, Northern Province, 14 May 1927, TNA 45/9, Accession No. 69.

23. Governor Cameron, quoted in CS to conservator of forests (CF), 7 June 1927, TNA 1733/A/13/I: 88.

24. For example, CF to CS, 4 April 1933, TNA 21559.

25. Troup, *Report on Forestry*, p. 35.

26. Kitching to CS, 23 May 1931, TNA H-11234.

27. D. J. J., comments on Executive Council Circular, 26 March 1933, TNA 2708.

28. CS to all PCs, 9 July 1931. Various PCs' replies to CS, 1931, TNA 20282.

29. Hoblen, secretary, SPFE, to SS, 31 July 1929, TNA 13582.

30. PC, Western Province, to CS, 10 December 1946, TNA 35773.

31. Comments on Forest Departmental Circular No. 1 of 1933, PC, Northern Province, n.d., TNA 21559.

32. Comments on Forest Departmental Circular No. 1 of 1933.

33. TT solicitor general's comments on the 1940 Game Ordinance,

Proceedings from the Legislative Council 14th Session, Part IV, May 1940, TNA 27273.

34. Tanzania was frequently singled out as the worst case in the empire because its laws were tolerant of African forms of hunting. East Africa High Commission, 1948, *Proceedings from a Conference Held on the Fauna of British Eastern and Central Africa in Nairobi, 8 and 9 May, 1947*, TNA 35773.

35. Appendix A, "The Game Position in Tanganyika Territory," in the *Proceedings of the 2nd African Game Conference*.

36. "Memorandum on Native Hunting," presented at the Conference Held on the Fauna of British Eastern and Central Africa in Nairobi.

37. M. Cowie, "Tanganyika: The Case for Preservation of Game," paper presented at the Conference Held on the Fauna of British Eastern and Central Africa in Nairobi, 8 and 9 May 1947, TNA 35773.

38. *London Times*, 18 February 1946. Another editorial charged that "native hunters are irresponsible and untrained in the humanities of hunting" and "[t]here can be no doubt that Tanganyika, among British territories, has the blackest record." *London Sunday Express*, 27 November 1929.

39. The SPFE singled out game drives and the use of pits as particularly deplorable. Hoblen, secretary, SPFE, to SS, 31 July 1929, TNA 13582.

40. M. Dunford, general manager, East African Tourist Travel Association to CS, 9 September 1949, TNA 37492.

41. Usa Planters Association to PC, Northern Province, 6 January 1930, TNA 13371.

42. Usa Planters Association to CS, 9 September 1931, TNA 13371.

43. Creech Jones, SS, to Gov. William Battershill, 10 June 1948, TNA 35773.

44. Member for Agriculture and Natural Resources to GW, 19 November 1949, TNA 37492.

45. PCs' comments on Game Department recommendations, 1941, TNA 13371.

46. Director of game preservation to CS, 12 January 1926, TNA 7623.

47. Acting GW to CS, 9 September 1991, TNA 35773.

48. CS to CF, 9 June 1931, TNA 10580.

49. Acting DC, same to PC, Tanga, 18 May 1934, TNA 13371.

50. PC, Southern Highlands Province, to Member for Agriculture and Natural Resources, 10 May 1948, TNA 37492.

51. Open letter from Capt. Hewlett, temporary game ranger, 5 September 1936, TNA 13371.

52. Comments on Executive Council Circular, initials illegible, 27 August 1938, TNA 13371.

53. PC, Lake Province, 19 April 1941, TNA Secretariat File 13371.

54. It is impossible to discern the accuracy of the reports of guards and rangers, for if the records reveal anything clearly, it is that their writings were consistently alarmist and exaggerated.

55. Game ranger, Bangai Hill, to PC, Lake Province, 25 July 1938, TNA 13371.

56. A game patrol arresting poachers in Serengeti near Musoma had to

retreat in the face of Ikoma and Sukuma hunting parties who wanted to attack them. "Report on Game Scout's Patrol in the Serengeti National Park" from Game Department, Bangai Hill, Musoma, 24 October 1946, TNA 35773.

57. The conservator mentioned that a "case of brutal assault has been reported recently in the Meru region." CF to CS, 21 November 1946, TNA 10948.

58. CF to CS, 7 April 1930, TNA 10948. CF to CS, 6 May 1937, TNA 24595.

59. Extract from the "Notes on the Discussion in the Finance Committee of the Legislative Council held at Arusha in December, 1929," TNA 10948.

60. Swynnerton, director of tsetse fly research, to CS, 25 April 1930, TNA 10948.

61. PC, Lake Province, to CS, 18 August 1938, TNA 13371.

62. CF to Member for Agriculture and Natural Resources, 3 October 1950, TNA 23185.

63. The following summary of honey gathering and Forest Department fire policy on Mount Meru is based partially on material in Neumann (1992).

64. CF to CS, 12 November 1947, TNA 10948.

65. PC, Mahenge, to CS, 11 June 1931, TNA 10948.

66. CF to CS, 14 March 1939, TNA 10948. This portion of the Meru Forest Reserve was officially closed on 10 May 1939, by Government Notice No. 66.

67. "Report on Fire Prevention," CF to CS, 21 November 1946, TNA 10948.

68. CF to Member for Agriculture and Natural Resources, 26 October 1948, TNA 10948.

69. The term "public lands" was adopted in the Tanganyika Territory Land Ordinance of 1923. It read, "The whole of the lands of the Territory, whether occupied or unoccupied, on the date of the commencement of this Ordinance are hereby declared to be public lands." In official correspondence and in many maps of the government survey office, the term "Crown lands" was often substituted. Resources, such as game and timber, were often said to be owned by the Crown.

70. The series begins with: Forester, Olmotonyi, to Arusha DC, 6 September 1926, TNA FOR/16, Accession No. 472. It concerned the harvesting of commercially valuable trees on Arusha and Meru native lands. In the following months the debate over timber royalties and ownership was taken to ever higher levels, eventually being decided in the CS's office.

71. Kitching, DC, Arusha, to CF, 9 September 1926, TNA FOR/16, Accession No. 472.

72. Kitching, DC, Arusha, to Mitchell, PC, Northern Province, 20 November 1926, TNA FOR/16, Accession No. 472.

73. Mitchell, PC, Northern Province, to CS, 29 March 1927, TNA FOR/16, Accession No. 472.

74. CF was quoted in a letter from the forester, Olmotonyi, to DO, Arusha, 4 October 1927, TNA FOR/16, Accession No. 472 (emphasis in the original).

75. Forester, Olmotonyi, to DO, Arusha, 14 September 1927, TNA FOR/16, Accession No. 472. As a point of clarification, the forester was quite wrong when he labeled African agricultural practices on Mount Meru "shifting cultivation." The Meru and Arusha had long been farming on fixed plots, and by the late 1920s intercropping with coffee had become widespread.

76. CF to DO, Arusha, 11 May 1928, TNA FOR/16, Accession No. 472.

77. Acting CS to CF, 2 May 1928, TNA FOR/16, Accession No. 472.

78. Ibid.

79. CF to Asst. CF, Moshi, 14 October 1930, TNA FOR/16, Accession No. 472.

80. Troup, *Report on Forestry.*

81. CF to CS, 19 November 1928, TNA 12913.

82. DO, Arusha, to PC, Northern Province, 9 February 1929, TNA 12913.

83. M. L. M. (unidentified administrator) to Member of Agriculture and Natural Resources, 6 July 1954, TNA 21559.

84. Forest surveyor to Northern Provincial Officer (PO), Arusha, 1 November 1955, TNA FOR/7, Accession No. 472.

85. Forester, Arusha District, to DC, Arusha, 14 April 1948, TNA FOR/7, Accession No. 472.

86. DC, Arusha, to Forester, Arusha District, 22 April 1948, TNA FOR/7, Accession No. 472.

87. Forester, Arusha District, to DC, Arusha, 5 May 1948, TNA FOR/7, Accession No. 472.

88. Forester, Arusha District, to DC, Arusha, 5 May 1948, TNA FOR/7, Accession No. 472.

89. Forester, Arusha District, to Asst. CF, 17 April 1948, TNA FOR/7, Accession No. 472.

90. Asst. CF (East Meru Charge) to DC, Arusha, 21 January 1960, TNA Arusha Regional File L/4/8/2.

91. Asst. CF (East Meru Charge) to DC, Arusha, 16 April 1960, TNA Arusha Regional File L/4/8/2.

92. Memo from DC, Arusha, to all *Hakimu,* 25 April 1960, TNA Arusha Regional File L/4/8/2.

Chapter Four

1. Masai of the national park (signed on behalf of Masai with thumbprint of Oltimbau ole Masiaya), 1957, "Memorandum on the Serengeti National Park" (5 pp.), RH, MSS. Afr. s. 1237b.

2. Swynnerton, director of game preservation, to CS, 17 January 1928, TNA 11234.

3. Amery, SS, to Gov. Cameron, 24 February 1928, TNA 12005.

4. SS to Gov. Symes, 14 February 1933, TNA 12005.

5. *Report of the Delegates of the International Congress for the Protection of Nature, Paris, June 1931, to His Majesty's Government,* TNA 12005.

6. Passfield to Gov. Cameron, 21 May 1930, TNA 19038.

7. Major Hingston, "Report on a Mission to East Africa," TNA 19038.

8. General Battye, GW, "Note on Major Hingston's Report," TNA 19038.

9. *Report of the Delegates of the International Congress for the Protection of Nature.*

10. The Ad Hoc Committee to Draft a Proposal for Action to be Debated by the International Convention, memorandum, TNA 12005.

11. Ibid.

12. Sir Cunliffe-Lister, SS, to Gov. Symes, confidential, 23 March 1933, with enclosed "Report of the Preparatory Committee for the International Conference for the Protection of the Flora and Fauna of Africa, 1933," TNA 12005.

13. Ibid.

14. Ibid.

15. Ibid.

16. "Summary of Observations on the Report of the Preparatory Committee," n.d., TNA 12005.

17. Acting Gov. Jardine to Cunliffe-Lister, SS, 1 August 1933, TNA 12005.

18. SS to Gov. MacMichael, 17 March 1934, TNA 12005.

19. "Note on the Convention," n.d., TNA 12005.

20. Gov. MacMichael to SS, 22 November 1937, TNA 12005.

21. D. M. K. to D. C. S., 17 February 1937, TNA 24979.

22. Memorandum, CS's office, 25 May 1933, TNA 11234.

23. [Illegible] to CS, 11 January 1938, TNA 12005.

24. Undersecretary of state to SPFE, 2 October 1939, TNA 12005.

25. "Comments on Major Hingston's Report on a Mission to East Africa," 9 April 1931, TNA 12005.

26. For example, the list of SPFE officers for 1933 read as follows: the patron was the Prince of Wales, the president was the Earl of Onslow, and the vice presidents were the Duke and Duchess of Bedford, Duke of Abercorn, Marquess of Crewe, Earl of Lonsdale, Viscount Grey, Viscount Allenby, Lord Lovat, Sir P. Chalmers Mitchell (SPFE 1933).

27. As a brief illustration, Onslow wrote to Malcolm MacDonald, secretary of state for the colonies, from his position in the House of Lords, explaining that they had not yet met, but that he knew his father well. He asked if they could meet to discuss "the preservation of fauna." Lord Onslow to MacDonald, 24 July 1935, PRO CO822|69/3. MacKenzie (1988, 286) provides a good example of the interrelations between imperial rule and conservation. The last three viceroys preceding WW II, Lord Irwin (later Lord Halifax, Onslow's brother-in-law), Willingdon, and Linlithgow, were hunters and conservationists. Willingdon's son became president of the Fauna Preservation Society (SPFE changed its name in 1950) and Linlithgow's father had led the British delegation in the 1900 international fauna conference.

28. As one example among many, Onslow reported to his fellow members "that he had put the question of National Parks down for discussion in Parliament" (SPFE 1930, 6).

29. Again, the examples are numerous. In an early instance, a delegation went to the secretary of state for the colonies to lobby for uniform laws and regulations for wildlife hunting and preservation (SPWFE, 1905). MacKenzie (1988) reviews some of these delegations in detail.

30. Onslow's appointments included: civil lord of the admiralty (1920–21); parliamentary secretary of the ministry of health (1921–23); and undersecretary of war (1925–26).

31. For example, the secretary of state for the colonies was anxious to get the colonial governors' endorsement of a coherent interterritorial wildlife protection policy in 1947 because he had "learnt that the Duke of Devonshire, who is President of the Society for the Preservation of the Fauna and Flora of the Empire, is proposing to initiate a debate on this subject in the House of Lords." CS of the Governors Conference to CS, TT, 26 August 1947, TNA 35773.

32. As MacKenzie (1987) pointed out, this idea could be carried to violent extremes. Quoting the writing of Baden-Powell he wrote, "In *Sport in War* he described the work involved in military operations as sporting itself, particularly the chase for 'wild beasts of the human kind,' which offered 'plenty of excitement and novel experience' among the kopjes and broken country of Rhodesia (Baden-Powell 1890). Elsewhere he remarked that 'the longest march seems short when one is hunting game . . . lion or leopard, boar or buck, nigger or nothing'" (MacKenzie 1987, 52).

33. SS to Gov. MacMichael, 25 March 1937, TNA 24979.

34. SS to Colonial Governments, Circular Despatch, 2 September 1938, TNA 12005.

35. *Report of the Special Committee Appointed to Examine the Game Bill, 1940*, 16 April 1940, TNA 27273.

36. Kitching, PC, Southern Province, to CS, 5 March 1940, TNA 27273.

37. SS to Gov. Battershill, 14 October 1946, TNA 35773.

38. Ibid.

39. CS, Tanganyika Territory, to CS, Governors Conference, 1 October 1947, TNA 35773.

40. SPFE secretary to undersecretary of state, 29 August 1939, TNA 12005.

41. Tanganyika Territory, National Parks Ordinance, 1948.

42. Memorandum No. 82 for Executive Council, 22 August 1950, TNA 34819.

43. P. Bleackley, secretary, Serengeti National Park Board of Trustees, to Member for Local Government, Dar es Salaam, 18 October 1951, TNA 10496.

44. Minutes of the second meeting of the Serengeti National Park Board of Trustees, 23 October 1951, TNA 10496.

45. Tanganyika Territory, "Proposals for Reconstituting the Serengeti National Park," Government Paper No. 5 of 1958, p. 3, University of Dar es Salaam Library, East Africana Collection.

46. Masai of the national park, 1957, "Memorandum on the Serengeti National Park."

47. Russel Arundel, "Petition to Alan Tindal Lennox-Boyd, Secretary of State for the Colonies," presented on behalf of the American Nature Conser-

vancy, National Parks Association, and Wilderness Society, among others. FPS Box Af/X1/NP.

48. Quoted in "Report from the Game Department on Game Scouts' Patrol in the National Park," 24 October 1946, TNA 35773.

49. PC, Western Province, to CS, 10 December 1946, TNA 35773.

50. Lt. Col. P. G. Molloy (director of national parks), "Native Poaching on the Western Serengeti Boundaries," memorandum. FPS Box Af/X1/NP.

51. DC, Masai/Monduli, to DC, Ngorongoro, 5 March 1955, TNA Arusha Regional File G1/6, Accession No. 69.

52. Lt. Col. P. G. Molloy (director of national parks), to DC, Masai/Monduli, 8 December 1955, TNA Arusha Regional File G1/6, Accession No. 69.

53. PC's notes on meeting with 110 Masai at Ngorongoro, 23 June 1954, TNA Arusha Regional File G1/6, Accession No. 69.

54. Masai of the national park, 1957, "Memorandum on the Serengeti National Park."

55. Ibid.

56. Acting PC, Northern Province, to Member for Local Government, 20 May 1955, TNA Arusha Regional File G1/6, Accession No. 69.

57. Masai of the national park, 1957, "Memorandum on the Serengeti National Park."

58. PC, Northern Province, to Member for Local Government 28 February 1955, TNA Arusha Regional File G1/6, Accession No. 69.

59. Major Hingston, "Report on a Mission to East Africa," TNA 19038.

60. PC, Northern Province, to Member for Local Government, 19 January 1955, TNA Arusha Regional File G1/6, Accession No. 69.

61. Notes on a meeting between the chairman of the National Park Board of Trustees, the director of National Parks, and the PC, Northern Province, 28 March 1955, TNA Arusha Regional File G1/6, Accession No. 69.

62. B. Leechman, chairman of the Serengeti National Park (SNP) Board of Management, from the minutes of the SNP Board of Management meeting, 23 July 1953, TNA 40851.

63. Coordinating Officer, Sukumaland Development, to PC, Lake Province, 5 May 1948, TNA 34819.

64. During hearings in the Legislative Council on the national park, one member remarked, "It will be a matter of great difficulty to establish rights, especially native rights, where there are no written records." Tanganyika Territory, *Proceedings from the Legislative Council,* 14th Session, May 1940, TNA 27273.

65. Notes on an informal discussion held at the Ngorongoro Rest Camp among members of the SNP Boards of Trustees and Management, TNA Arusha Regional File T3/2, Accession No. 69. The notes state that the park administration "would be both willing and able to evict all non-Masai from the park within a year or two."

66. DC, Masai/Monduli, to PC, Northern Province, 23 June 1952, TNA Arusha Regional File T3/2, Accession No. 69.

67. PC, Northern Province, "Memorandum on Serengeti Evictions," 19 January 1955, TNA Arusha Regional File G1/6, Accession No. 69.

68. PC, Northern Province, "Memorandum on Serengeti evictions," 19 January 1955, TNA Arusha Regional File G1/6, Accession No. 69.

69. Gov. Twining, Tanganyika, to Gov. Baring, Kenya, 15 December 1953, PRO CO 822/502.

70. Ibid.

71. Extract from the Political Intelligence Summary for February 1955, PRO CO 822/806.

72. Molloy, director, Tanzania National Parks (TANAPA), to PC, Northern Province, "Report on Human Inhabitants, Serengeti National Park," 8 June 1955, TNA Arusha Regional File G1/6, Accession No. 69.

73. Wilkins, SNP Board of Management, to SNP Board of Trustees, 16 February 1954, TNA 10496.

74. Gov. Twining to the SNP Board of Trustees, 16 November 1953, TNA 10496.

75. The governor had his hands full keeping Mau Mau out of Tanganyika, and he wanted to avoid additional trouble with the Maasai. A month after issuing his threat to decommission Serengeti National Park, Twining ordered "Operation Mistletoe," a Christmas Eve raid on eighteen locations to "smash the militant Mau Mau organizations in our Northern Province." Gov. Twining, Tanganyika, to Gov. Baring, Kenya, 15 December 1953, PRO CO 822/502.

76. Acting PC, Northern Province, to director of national parks, 8 June 1955, TNA Arusha Regional File G1/6, Accession No. 69. The PC pointed out that a recent census had determined that 82 out of 216 families cultivating in the crater were Maasai.

77. Molloy, director, TANAPA, to PC, Northern Province, "Report on Human Inhabitants, Serengeti National Park," 8 June 1955, TNA Arusha Regional File G1/6, Accession No. 69.

78. Notes on an informal discussion held at the Ngorongoro Rest Camp among members of the SNP Boards of Trustees and Management, TNA Arusha Regional File T3/2, Accession No. 69.

79. DC, Masai/Monduli, to PC, Northern Province, 23 June 1952, TNA Arusha Regional File T3/2, Accession No. 69.

80. Tanganyika Territory, "Report of the Serengeti Committee of Inquiry," 1957, University of Dar es Salaam Library, East Africana Collection.

81. "The Serengeti National Park," Tanganyika Sessional Paper No. 1 of 1956, University of Dar es Salaam Library, East Africana Collection.

82. "Petition to Alan Tindal Lennox-Boyd, SS for the Colonies, Presented by Russell M. Arundel in Behalf of the American Nature Conservancy, American Nature Association, National Parks Association, American Committee for International Wildlife Protection, Wildlife Management Institute, National Wildlife Federation, and the Wilderness Society," n.d., p. 6. FPS Archives AF/X1/NP.

83. Tanganyika Territory, "Report of the Serengeti Committee of Inquiry."

84. "Proposals for Reconstituting the Serengeti National Park," Tanganyika Sessional Paper No. 5 of 1958, University of Dar es Salaam Library, East Africana Collection.

85. TANAPA, *Report and Accounts of the Board of Trustees,* July 1959 to June 1960.

86. Address delivered by His Excellency Sir Richard Turnbull, Governor of Tanganyika, 5 September 1961, at the opening session, CCTA/IUCN Symposium on the Conservation of Nature and Natural Resources in Modern African States.

87. Ibid.

88. TANAPA, *Report and Accounts of the Board of Trustees,* 1959–60.

89. The film was partly funded by the New York Zoological Society. TANAPA, *Report and Accounts of the Board of Trustees,* 1961–62.

90. TANAPA, *Report and Accounts of the Board of Trustees,* 1960–61.

91. TANAPA, Board of Trustees, *Annual Report,* 1966–67.

92. Minutes of a meeting of the TANAPA Board of Trustee's Sub-Committee on Localization, 7 February 1970, TANAPA Closed Files (housed at Arusha National Park).

93. "Tanzania National Parks Management Problems," paper prepared by the Ministry of Information and Tourism, ca. October 1970, TANAPA Closed Files (housed at Arusha National Park).

94. At the time of this study, the director and assistant director of TANAPA were Mweka classmates, and nearly every warden in the field was a Mweka graduate.

95. TANAPA, Board of Trustees, *Annual Report,* 1968–69.

96. Anonymous memo concerning a 1970 study on the development of the tourist industry by Arthur D. Little, Inc., TANAPA Closed Files (housed at Arusha National Park).

97. The governor complained about it in a memorandum to his staff, suggesting ways to force tourists to spend money in the country. Gov. of TT, memorandum, 2 February 1935, TNA 22856.

98. In July of 1988, during the operation to relocate the pastoralists, I visited the reserve in the company of the officer in charge. Although the operation had been completed just prior to my arrival, arrests were still being made of people trying to reenter the reserve. Many of the former residents refused to cooperate and declined the offer of the government to transport them to their new settlement area.

99. General Battye, GW, "Note on Major Hingston's Report," TNA 19038.

100. TANAPA, *Report and Accounts of the Board of Trustees,* 1961–62.

101. "Tanganyika Territory, 1960 Council Debates, Tanganyika Legislative Council Official Report, 35th Session, 26 to 27 April 1960," University of Dar es Salaam, East Africana Collection.

102. Conversation with former Meru Mangi (Chief) Sylvanus Kaaya, 14 August 1990. Although encompassed within the lands claimed by the Meru, the swampy crater was not integral to their land use practices.

103. TANAPA, *Report and Accounts of the Board of Trustees,* 1961–62.

104. TANAPA, *Report and Accounts of the Board of Trustees,* 1962–64.

105. The Meru Land Committee's agreement is discussed by acting permanent secretary to DC, Arusha, 14 June 1962, and acting DC, Arusha, to

director, TANAPA, 3 January 1962, Vesey-Fitzgerald Papers, Serengeti Wildlife Research Institute (SWRI) Library. The Meru's agreement to give up a piece of land to the park can be understood in terms of what was to be gained and lost in the trade-off. First, Meru herders did not water their livestock at Momella Lakes, since the waters are alkaline. In any case, perhaps 90 percent of the lakes' shoreline was already protected within the park and off-limits to grazing, and the land area to be added to the park was quite small. Second, and perhaps most important, the Meru would be given proper title—providing a security of tenure that they had never had—to by far the greater portion of the former public land.

106. TANAPA, *Report and Accounts of the Board of Trustees, 1962–64*.

107. TANAPA, *Director's Annual Report, 1969–70*.

108. DC, Arusha, to PC, Northern Province, 1 March 1956, TNA FOR/7, Accession No. 472.

109. Director, TANAPA to TANAPA Board of Trustees (draft memorandum), n.d., ca. 1961, Records of the Arusha National Park (ANP) Headquarters.

110. Village chairman, Nkoasenga, to director, TANAPA, 12 October 1982, Records of the ANP Headquarters.

111. ANP warden to director, TANAPA, 20 October 1977, Records of the ANP Headquarters. Original in Swahili; translation by Alawi Msuya.

112. Handwritten response by the director on above memo, Records of the ANP Headquarters. Original in Swahili; translation by Alawi Msuya.

113. Interviews with anonymous Nasula residents, July and August 1990.

114. Interview with Sori Kibata [pseud.], 18 July 1990.

115. In this particular passage—which concerned the establishment of Lake Manyara National Park—the validity of preexisting claims to land and resources is further qualified by the author's use of quotation marks around the term "rights." TANAPA, *Report and Accounts of the Board of Trustees, 1959–60*.

116. TANAPA Board of Trustees chairman, from the foreword to TANAPA, Board of Trustees, *Annual Report, 1967–68*.

117. TANAPA, *Report and Accounts of the Board of Trustees, 1962–64*.

Chapter Five

1. From a series of interviews conducted with Meru elders between January 1990 and August 1990.

2. Ibid.

3. TANAPA, *Report and Accounts of the Board of Trustees, 1961–62*.

4. D. F. Vesey-FitzGerald, "Integration Report," for the Tanzania National Parks Ecological Unit to the Serengeti Research Institute, 1973, TS, Vesey-FitzGerald Papers, SWRI Library.

5. This is a management practice begun in the early days of the park but neglected during the 1980s, probably because of the lack of staff and funds to do the work. Under the park warden who was in charge during my research, the program was begun again in January of 1990 using casual labor hired in the sur-

rounding villages. The brush is cut using *pangas*, then left to dry and later burned.

6. *Arusha National Park Monthly Report*, January 1987, Records of the ANP Headquarters.

7. *Arusha National Park Monthly Report*, October 1986, Records of the ANP Headquarters.

8. *Arusha National Park Monthly Report*, June 1986, Records of the ANP Headquarters.

9. Summarized from various *Arusha National Park Monthly Reports*, including May 1983, September 1985, and June 1986, Records of the ANP Headquarters.

10. *Arusha National Park Monthly Report*, October 1977, December 1977, May 1978, and May 1979; *Kusare Guard Post Report*, March 1977, Records of the ANP Headquarters.

11. *Arusha National Park Monthly Report*, May 1978 and May 1979, Records of the ANP Headquarters.

12. *Arusha National Park Monthly Report*, May 1979, Records of the ANP Headquarters.

13. *Rydon Guard Post Report*, January–April 1979, Records of the ANP Headquarters.

14. Interviews with Nasula and Ngongongare farmers, 14, 20, and 28 July 1990.

15. This profile is constructed from park arrest reports and interviews with various park staff members.

16. *Arusha National Park Monthly Report*, July 1985, Records of the ANP Headquarters.

17. *Arusha National Park Monthly Report*, November 1975, Records of the ANP Headquarters.

18. *Arusha National Park Monthly Report*, April 1990, Records of the ANP Headquarters.

19. *Arusha National Park Monthly Report*, March 1990, Records of the ANP Headquarters.

20. *Internal Report*, chief park warden, ANP to Director, TANAPA, 21 May 1976, Records of the ANP Headquarters.

21. *Arusha National Park Monthly Report*, January 1974, Records of the ANP Headquarters.

22. *Arusha National Park Monthly Report*, March 1974, Records of the ANP Headquarters.

23. D. F. Vesey-FitzGerald to M. Turner, ANP Internal Memo, 6 November, 1973, Records of the ANP Headquarters.

24. *Arusha National Park Monthly Report*, March 1980, Records of the ANP Headquarters.

25. From various interviews with villagers in Nasula in March 1990 and Ngongongare in July 1990.

26. Another TANAPA effort to reduce the level of local resistance at Arusha National Park was implemented in 1991, after my study there was completed. As part of the agency's new Community Conservation Service,

developed with the help of AWF, a special Community Conservation Warden (CCW) was assigned to Arusha National Park (see Bergin 1995). The principal role of the CCW is to work on management issues beyond the park boundary.

27. *Arusha National Park Monthly Report,* March 1977, Records of the ANP Headquarters.

28. *Arusha National Park Monthly Report,* October 1989, Records of the ANP Headquarters.

29. *Arusha National Park Monthly Report,* April 1974, Records of the ANP Headquarters.

30. From the government's obsolete "Arusha National Park Development Plan," ca. 1981, Records of the ANP Headquarters.

31. Interview with Kulika [pseud.], Ngongongare, 2 July 1990.

32. Interview with Ishmail [pseud.], Nasula, 18 April 1990.

33. D. F. Vesey-FitzGerald, "Ngurdoto National Park Electric Fence," 1967, TS, Vesey-FitzGerald Papers, SWRI Library.

34. TANAPA, *Report and Accounts of the Board of Trustees, 1964–67.*

35. *Arusha National Park Monthly Report,* August 1988, Records of the ANP Headquarters.

36. I encountered two documented incidences of park rangers assisting villagers in crop defense, in the *Arusha National Park Monthly Report,* August 1986, and the *Kinandia Guard Post Report,* June 1977, Records of the ANP Headquarters.

37. Game management officer, Game Division, to Asst. Park Warden, ANP, 30 June 1966, Records of the ANP Headquarters.

38. Director, TANAPA, to Game Division, 11 July 1966, Records of the ANP Headquarters.

39. Chief Park Warden, ANP, to Game Division, 20 April 1989, Records of the ANP Headquarters.

40. Chief Park Warden, ANP, to village chairman, Nkoasenga, 22 May 1985, Records of the ANP Headquarters.

41. Chief Park Warden, ANP, to village chairman, Nkoasenga, 29 May 1985, Records of the ANP Headquarters.

42. Interview with Mzee Saliwa [pseud.], 9 July 1990.

43. Chief Park Warden, ANP, to Director, TANAPA, 14 August 1973, Records of the ANP Headquarters.

Chapter Six

1. The linkage of land rights to social identity is strengthened under situations such as existed on Mount Meru during the colonial period. That is, the British mapped out lands reserved for the exclusive use of the Meru "tribe." Under these circumstances, throughout British colonial Africa, land rights were welded to social identity (Berry 1992).

2. Similar themes of contested histories and myths of park landscapes are developed by Moore (1993) and Ranger (n.d.) for Zimbabwe. Moore explains that the state's view of Nyanga National Park as a commodity and source of revenue contrasts sharply with local meanings that recognize the "sacred features

of the landscape" (1993, 391). Ranger, in relating the complex and dynamic web of meanings attached to the Matopos National Park, shows how whites have historicized the landscape with heroic European myths. This landscape overlaps, and contrasts with, Karanga history, which reads in the Matopos Hills its myth of creation.

3. Village chairman, Nkoasenga, to director, TANAPA, 12 October 1982, Records of the ANP Headquarters.

4. Interview with Seneto Village chairman, 12 September 1990.

5. *Monthly Report of the Arusha National Park Warden*, January 1974, Records of the ANP Headquarters.

6. *Monthly Report of the Arusha National Park Warden*, February 1974, Records of the ANP Headquarters.

7. *Monthly Report of the Arusha National Park Warden*, March 1974, Records of the ANP Headquarters.

8. Interview with Yuri Mtabu [pseud.], 4 September 1990. A ten-cell leader is a member of the ruling party, Chama Cha Mapanduzi, who is responsible for organizing a cell of ten households.

9. Interview with Paramikya Mbise [pseud.], 22 January 1990.

10. The "permanent crops" refer to banana and papaya trees mostly, though some people with whom I spoke talked of wanting to plant coffee. "Minutes of the Nasula Kitongoji Committee," June 1989. Translation from the Swahili original by Alawi Msuya.

11. Interview with David Sarkikiya [pseud.], 21 April 1990.

12. Interview with Peter Karine [pseud.], 1 March 1990.

13. *Monthly Report of the Arusha National Park Warden,* October 1974, Records of the ANP Headquarters.

14. *Headquarters Ranger Post Report,* March 1974, Records of the ANP Headquarters.

15. CF to CS, 19 November 1928, TNA 12913.

16. *Arusha National Park Warden Monthly Report,* August 1980, Records of the ANP Headquarters.

17. *Arusha National Park Warden Monthly Report,* January 1975, Records of the ANP Headquarters.

18. Interview with Karika Mkuti [pseud.], 19 June 1990.

19. Interview with Urisho Mbati [pseud.], 12 April 1990.

20. Interview with Peter Mkikya [pseud.], 25 June 1990.

21. Interview with Joseph Kitanda [pseud.], 20 January 1990.

22. Interview with Mswana Pilau [pseud.], 19 April 1990.

23. Interview with Mswana Pilau [pseud.], 1990.

24. Interview with Sori Kibata, 6 May 1990.

25. Interview with Isiah Mukuti [pseud.], 7 July 1990.

26. This is only one aspect of a socially complex and dynamic relationship between park staff and villagers. There are other more cooperative and mutually beneficial aspects, which I will discuss below. It is also an accusation that only village men made to me. No women told me of being the actual victims, but I would not expect it to be otherwise. Some of the men explained it as simply typical behavior on the part of police and military personnel.

27. Interview with Duluti Kikula [pseud.], 12 June 1990.

28. Interview with Ngurdoto resident, 11 September 1990.

29. Interview with Duluti Kikula.

30. Ibid.

31. Interview with Joseph Kitanda.

32. Interview with William Mbili [pseud.], 6 June 1990.

33. From various interviews in Ngongongare and Nasula, April–July 1990.

34. Interview with Josephine Likula [pseud.], 11 June 1990.

35. Interview with Doris Karika [pseud.], 14 April 1990.

36. Interview with Martin Karika [pseud.], 9 July 1990.

37. Interview with Joseph Matabu [pseud.], 18 April 1990.

38. Interview with Mswana Pilau.

39. Interview with Isiah Mbise [pseud.], 25 January 1990.

40. Interview with Dembe Urio [pseud.], 12 June 1990.

41. Ibid.

42. Interview with Emanuel Mkuti [pseud.], 24 January 1990.

43. Interview with Dembe Urio.

44. Interview with Seneto Village chairman.

45. Interview with Mary Mkituni [pseud.], 16 March 1990.

46. Interview with Mswana Pilau.

47. Ibid.

48. Interview with Martin Karika.

49. In April 1983 the warden wrote, "Due to lack of funds, staff are going hungry since most of them are with their families here. It appears to the workers, especially the rangers, that they should stop working." *Monthly Report of the Arusha National Park Warden*, March 1974; April and May 1983, Records of the ANP Headquarters.

50. Interview with Emanuel Mkuti.

51. Interview with Josephine Likula.

52. Interview with Emanuel Tatuni [pseud.], 15 April 1990.

53. Interview with Karika Mkuti [pseud.], 19 June 1990.

54. Interview with Martin Karika.

55. TANAPA, *Annual Report 1985–86*.

56. TANAPA, *Arusha National Park Master Plan*, produced by the College of African Wildlife Management Diploma Class, 1980, Records of the ANP Headquarters.

57. *Arusha National Park Warden Monthly Report*, July 1974, Records of the ANP Headquarters.

58. *Arusha National Park Warden Monthly Report*, August 1980, Records of the ANP Headquarters.

59. *Arusha National Park Warden Monthly Report*, January 1975, Records of the ANP Headquarters.

60. *Arusha National Park Warden Monthly Report*, March 1975, Records of the ANP Headquarters.

61. Interview with Doris Karika.

62. Interview with Francis Ishirini [pseud.], 28 March 1990.

63. Interview with Dembe Urio.

64. Interview with David Sarkikiya.

Literature Cited

Documentary Sources

East Africana Collection, University of Dar es Salaam Library, Tanzania
Fauna Preservation Society Archives (FPS), London Zoological Society,
  Regents Park, UK
Onslow Family Papers, Surrey Record Office, Guildford, UK
Public Record Office (PRO), Kew, UK
Rhodes House Library (RH), Oxford, UK
Serengeti Wildlife Research Institute (SWRI) Library, Seronera, Tanzania
Tanzania National Archives (TNA), Dar es Salaam, Tanzania

Published Sources

Adams, Jonathan S., and T. O. McShane. 1992. *The myth of wild Africa: Con-
  servation without illusion.* New York: W. W. Norton and Company.
Akram-Lodhi, A. Haroon. 1992. "Peasants and hegemony in the work of
  James Scott." *Journal of Peasant Studies* 19 (3–4): 179–201.
Ames, Evelyn. 1967. *A glimpse of Eden.* Boston: Houghton Mifflin Company.
Anderson, David. 1986. "Stock theft and moral economy in colonial Kenya."
  Africa 56 (4): 399–416.
Anderson, D., and R. Grove. 1987. "Introduction: The scramble for Eden:
  Past, present and future in African conservation." In *Conservation in
  Africa: People, policies and practice,* edited by D. Anderson and R. Grove,
  1–12. Cambridge: Cambridge University Press.
Arhem, Kaj. 1984. "Two sides of development: Maasai pastoralism and
  wildlife conservation in Ngorongoro, Tanzania." *Ethnos* 49 (3–4):
  186–210.

————. 1985. *Pastoral man in the Garden of Eden: The Maasai of the Ngorongoro Conservation Area, Tanzania*. Uppsala: University of Uppsala.

Arnold, D. 1984. "Gramsci and peasant subalternality." *Journal of Peasant Studies* 11:155–77.

Badshah, M. A., and C. A. R. Bhadran. 1962. "National parks: Their principles and purposes." In *First World Conference on National Parks*, edited by A. B. Adams, 23–24. Washington D.C.: U.S. Government Printing Office.

Bailey, F. G. 1987. "The Peasant View of the Bad Life." In *Peasants and Peasant Societies*, edited by Teodor Shanin, 284–299. 2nd ed. New York: Basil Blackwell.

Barrell, John. 1980. *The dark side of the landscape: The rural poor in English painting 1730–1840*. Cambridge: Cambridge University Press.

Barrett, Christopher B., and Peter Arcese. 1995. "Are integrated conservation-development projects (ICDPs) sustainable? On the conservation of large mammals in sub-Saharan Africa." *World Development:* 23 (7): 1073–84.

Baskin, Yvonne. 1994. "There's a new wildlife policy in Kenya: Use it or lose it." *Science* 265:733–34.

Beinart, William. 1984. "Soil erosion, conservationism, and ideas about development: A Southern African exploration, 1900–1960." *Journal of Southern African Studies* 11 (1): 52–83.

————. 1989. "Introduction: The politics of colonial conservation." *Journal of Southern African Studies* 15 (2): 143–62.

Beinart, W., and C. Bundy. 1987. *Hidden struggles in rural South Africa: Politics and popular movements in the Tanskei and Eastern Cape*. Berkeley and Los Angeles: University of California Press.

Berger, John. 1980. *About looking*. London: Writers and Readers.

Bergin, Patrick. 1995. *Conservation and Development: The institutionalization of community conservation in Tanzania National Parks*. Ph.D. diss., University of East Anglia.

Bernstein, Henry. 1988. "Capitalism and petty-bourgeois production: Class relations and divisions of labour." *Journal of Peasant Studies* 15 (1): 258–71.

Berry, Sara. 1989. "Social institutions and access to resources." *Africa* 59 (1): 41–55.

————. 1992. "Hegemony on a shoestring: Indirect rule and access to agricultural land." *Africa* 62 (3): 327–55.

Bienen, Henry. 1970. *Tanzania: Party transformation and economic development*. Princeton: Princeton University Press.

Bird, Cherry, and Simon Metcalfe. 1995. *Two views from CAMPFIRE in Zimbabwe's Hurungwe District: Training and motivation. Who benefits and who doesn't?* IIED Wildlife and Development Series Number 5. London: International Institute for Environment and Development.

Bird, Elizabeth Ann. 1987. "The social construction of nature: Theoretical approaches to the history of environmental problems." *Environmental Review* 11 (4): 255–64.

Blaikie, Piers. 1985. *The political economy of soil erosion in developing countries*. London: Longman.

Blaikie, Piers, and Harold Brookfield. 1987. *Land degradation and society.* London: Methuen.

Blaut, J. M. 1993. *The colonizer's model of the world: Geographical diffusionism and Eurocentric history.* New York: Guilford Press.

Bloch, P. 1993. *Buffer zones, buffering strategies, resource tenure, and human-natural resource interaction in the peripheral zones of protected areas in sub-Saharan Africa.* Madison: Land Tenure Center.

Bolton, Dianne. 1985. *Nationalization—a road to socialism?: The lessons of Tanzania.* London: Zed Books.

Bonner, Raymond. 1993. *At the hand of man: Peril and hope for Africa's wildlife.* New York: Alfred A. Knopf.

Borner, Markus. 1985. "The increasing isolation of Tarangire National Park." *Oryx* 19 (2): 91–96.

Bruce, John. 1993. "Do indigenous tenure systems constrain agricultural development?" In *Land in African agrarian systems,* edited by Thomas J. Bassett and Donald E. Crummey, 35–56. Madison: University of Wisconsin Press.

Bruce, Michael. 1994. *The countryside ideal: Anglo-American images of landscape.* London: Routledge.

Bunn, David. 1994. "'Our wattled cot': Mercantile and domestic space in Thomas Pringle's African landscapes." In *Landscape and power,* edited by W. J. T. Mitchell, 127–74. Chicago: University of Chicago Press.

Burgess, Rod. 1978. "The concept of Nature in geography and Marxism." *Antipode* 10 (2): 1–11.

Cameron, Sir Donald. 1939. *My Tanganyika service, and some Nigeria.* 2nd ed. Washington, D.C.: University Press of America.

Carney, J. 1993. "Converting the wetlands, engendering the environment: The intersection of gender with agrarian change in the Gambia." *Economic Geography* 69 (4): 329–48.

Carney, J., and M. Watts. 1990. "Manufacturing dissent: Work, gender and the politics of meaning in a peasant society." *Africa* 60 (2): 207–41.

Caruthers, Jane. 1989. "Creating a national park, 1910 to 1926." *Journal of Southern African Studies* 15 (2): 188–216.

———. 1994. "Dissecting the myth: Paul Kruger and the Kruger National Park." *Journal of Southern African Studies* 20 (2): 263–83.

Chase, Alston. 1987. *Playing God in Yellowstone: The destruction of America's first national park.* New York: Harcourt, Brace, Jovanovich.

Child, Graham. 1996. "The role of community-based wild resource management in Zimbabwe." *Biodiversity and Conservation* 5:355–67.

Clark, T. J. 1984. *The painting of modern life: Paris in the art of Manet and his followers.* New York: Alfred A. Knopf.

Cleaver, Kevin. 1993. *A strategy to develop agriculture in sub-Saharan Africa and a focus for the World Bank.* World Bank Technical Paper Number 203. Washington, D.C.: World Bank.

Colchester, Marcus. 1994. Salvaging nature: Indigenous peoples, protected areas, and biodiversity conservation. UNRISD Discussion Paper. Geneva: United Nations Research Institute for Social Development.

Collett, David. 1987. "Pastoralists and wildlife: Image and reality in Kenya

Maasailand." In *Conservation in Africa: People, policies and practice,* edited by David Anderson and Richard Grove, 129–48. Cambridge and New York: Cambridge University Press.

Cooper, Frederick. 1994. "Conflict and connection: Rethinking colonial African history." *American Historical Review* 99 (5): 1516–45.

Cosgrove, Denis E. 1984. *Social formation and symbolic landscape.* London: Croom Helm.

———. 1995. "Habitable Earth: Wilderness, empire, and race in America." In *Wild ideas,* edited by David Rothenburg, 27–41. Minneapolis: University of Minnesota Press.

Cosgrove, Denis, and Stephen Daniels, eds. 1988. *The iconography of landscape: Essays on the symbolic representation, design, and use of past environments.* Cambridge: Cambridge University Press.

Coulson, Elizabeth. 1971. "The impact of the colonial period on the definition of land rights." In *Colonialism in Africa 1870–1960,* edited by Victor Turner, 193–215. Cambridge: Cambridge University Press.

Crandell, Gina. 1993. *Nature pictorialized: "The view" in landscape history.* Baltimore: John Hopkins University Press.

Cronon, William. 1983. *Changes in the land: Indians, colonists, and the ecology of New England.* New York: Hill and Wang.

Crosby, Alfred W. 1986. *Ecological imperialism: The biological expansion of Europe, 900–1900.* Cambridge: Cambridge University Press.

Crummey, Donald, ed. 1986. *Banditry, rebellion and social protest in Africa.* London: James Currey.

Curry, Steve. 1982. *Wildlife conservation and game-viewing tourism in Tanzania.* ERB Paper 82.5. Dar es Salaam: Economic Research Bureau, University of Dar es Salaam.

Curry-Lindahl, K. 1963. "Developing an appreciation for the need to conserve nature and natural resources." In *Conservation of nature and natural resources in modern African states,* edited by Gerald G. Watterson, 54–59. Morges, Switzerland: International Union for the Conservation of Nature and Natural Resources.

Curtin, Philip D. 1964. *The image of Africa: British ideas and action, 1780–1850.* Madison: University of Wisconsin Press.

DANIDA. 1989. *Environmental profile: Tanzania.* Copenhagen: Danish International Development Agency.

Daniels, Stephen. 1993. *Fields of vision: Landscape imagery and national identity in England and the United States.* Cambridge: Polity Press.

Daniels, Stephen, and Denis Cosgrove. 1988. "Introduction: Iconography and landscape." In *The iconography of landscape: Essays on the symbolic representation, design, and use of past environments,* edited by Denis E. Cosgrove and Stephen Daniels, 1–10. Cambridge: Cambridge University Press.

Dasmann, Raymond R. 1984. "The relationship between protected areas and indigenous peoples." In *National parks, conservation and development,* edited by J. McNeely and K. Miller, 667–71. Washington: Smithsonian Institution Press.

Demeritt, David. 1994. "The nature of metaphors in cultural geography and environmental history." *Progress in Human Geography* 18 (2): 163–85.

Diehl, C. 1985. "Wildlife and the Maasai: The story of East African parks." *Cultural Survival Quarterly* 9 (1): 37–40.

EIU. 1996. *Economist Intelligence Unit Country Report, Tanzania, 1st Quarter.* London: Economist Intelligence Unit.

Ellis, Stephen. 1994. "Of elephants and men: Politics and nature conservation in South Africa." *Journal of Southern African Studies* 20 (1): 53–69.

Evans, Grant. 1987. "Sources of peasant consciousness in South-East Asia: A survey." *Social History* 12 (2): 193–211.

Feierman, Steven. 1990. *Peasant intellectuals: Anthropology and history in Tanzania.* Madison: University of Wisconsin Press.

Fitter, R., and Sir Peter Scott. 1978. *The penitent butchers: The Fauna Preservation Society, 1903–1978.* London: Zoological Society of London.

FitzSimmons, Margaret. 1989. "The matter of nature." *Antipode* 21 (2): 106–20.

Foner, Eric, and Jon Wiener. 1991. "Fighting for the West." *Nation* (29 July–5 August): 163–66.

Fortmann, Louise. 1980. *Peasants, officials and participation in rural Tanzania: Experience with villagization and decentralization.* Ithaca: Rural Development Committee, Cornell University.

Fosbrooke, Henry. 1990. "Pastoralism and land tenure." Paper presented at the Workshop on Pastoralism and the Environment, April 1990, Arusha, Tanzania.

Frykman, Jonas, and Orvar Lofgren. 1987. *Culture builders: A historical anthropology of middle-class life.* New Brunswick: Rutgers University Press.

Geertz, Clifford. 1973. "Thick description: Toward an interpretive theory of culture." In *The interpretation of cultures: Selected essays,* 3–33. New York: Basic Books.

Ghai, Dharam. 1992. *Conservation, livelihood, and democracy: Social dynamics of environmental changes in Africa.* UNRISD Discussion Paper Number 33. Geneva: United Nations Research Institute for Social Development.

Ghimire, Krishna. 1994. "Parks and people: Livelihood issues in national park management in Thailand and Madagascar." *Development and Change* 25:195–229.

Gibson, C. C., and S. A. Marks. 1995. "Transforming rural hunters into conservationists: An assessment of community-based wildlife management programs in Africa." *World Development* 23 (6): 941–57.

Glacken, Clarence J. 1967. *Traces on the Rhodian shore: Nature and culture in Western thought from ancient times to the end of the eighteenth century.* Berkeley and Los Angeles: University of California Press.

Goheen, Mitzi. 1992. "Chiefs, sub-chiefs, and local control: Negotiations over land, struggles over meaning." *Africa* 62 (3): 389–412.

Gordon, Robert. 1992. *The Bushman myth: The making of a Namibian underclass.* Boulder: Westview.

Gramsci, Antonio. 1971. *Selections from the prison notebooks.* Edited and

Translated by Quintin Hoare and Geoffrey N. Smith. New York: International Publishers.

Grove, Richard. 1990. "Colonial conservation, ecological hegemony and popular resistance: Toward a global synthesis." In *Imperialism and the natural world,* John MacKenzie, 1–14. Manchester: Manchester University Press.

Harmon, David. 1987. "Cultural diversity, human subsistence, and the national park ideal." *Environmental Ethics* 9 (summer): 147–58.

Hay, Douglas. 1975. "Poaching and the game laws on Cannock Chase." In *Albion's fatal tree: Crime and society in eighteenth century England,* edited by Douglas Hay, Peter Linebaugh, John Rule, E. P. Thompson, and Cal Winslow, 189–254. New York: Pantheon Books.

Hecht, Susanna, and Alexander Cockburn. 1990. *The fate of the forest: Developers, destroyers and defenders of the Amazon.* New York: HarperCollins.

Hill, Kevin. 1996. "Zimbabwe's wildlife utilization programs: grassroots democracy or an extension of state power?" *African Studies Review* 39 (1): 103–22.

Hobsbawm, Eric J. 1985. *Bandits.* 2nd ed. Harmondsworth: Penguin.

Homewood, K. M., and W. A. Rodgers. 1984. "Pastoralism and conservation." *Human Ecology* 12 (4): 431–41.

Hopkins, Harry. 1985. *The long affray: The poaching wars, 1760–1914.* London: Secker and Warburg.

Huxley, Julian. 1961. *The conservation of wild life and natural habitats in central and East Africa: Report on a mission accomplished for UNESCO.* Paris: UNESCO.

Hyden, Goran. 1980. *Beyond Ujamaa in Tanzania: Underdevelopment and an uncaptured peasantry.* Berkeley and Los Angeles: University of California Press.

Iliffe, John. 1969. *Tanganyika under German rule, 1905–1912.* Cambridge: Cambridge University Press.

———. 1979. *A modern history of Tanganyika.* Cambridge: Cambridge University Press.

Isaacman, Allen. 1990. "Peasants and rural social protest in Africa." *African Studies Review* 33 (2): 1–120.

Isaacman, A., M. Stephen, Y. Adam, M. J. Homen, E. Macamo, and A. Pililao. 1980. "Cotton is the mother of poverty": Peasant resistance to forced cotton production in Mozambique." *International Journal of African Historical Studies* 13 (4): 581–615.

IUCN. 1987. *Directory of Afrotropical Protected Areas.* Gland, Switzerland: International Union for the Conservation of Nature and Natural Resources.

———. 1991. *Protected areas of the world: A review of national systems.* Vol. 3 of Afrotropical. Gland, Switzerland: World Conservation Union.

Jackson, Peter. 1989. *Maps of meaning: An introduction to cultural geography.* London: Routledge.

James, R. W. 1971. *Land tenure and policy in Tanzania.* Toronto: University of Toronto Press.

Japhet, K., and E. Seaton. 1967. *The Meru land case.* Nairobi: East African Publishing House.

Kelso, Casey. 1993. "The landless Bushmen." *Africa Report* 38 (2): 51–54.

KIPOC. 1992. *The Foundation program: Program profile and rationale.* Principal Document Number 4. Loliondo, Tanzania: Korongoro Integrated Peoples Oriented to Conservation.

Kiss, A. 1990. *Living with wildlife: Wildlife resource management with local participation in Africa.* World Bank Technical Paper Number 130. Washington: World Bank.

Kjekshus, H. 1977. *Ecology control and economic development in East African history: The case of Tanganyika, 1850–1950.* Berkeley and Los Angeles: University of California Press.

Koponen, Juhani. 1988. *People and production in late precolonial Tanzania: History and structures.* Finnish Society for Development Studies, Monograph Number 2. Helsinki.

———. 1995. *Development for exploitation: German colonial policies in mainland Tanzania, 1884–1914.* Helsinki: Tiedekirja.

Lamprey, Hugh. 1992. "Challenges facing protected area management in sub-Saharan Africa." In *Managing protected areas in Africa,* edited by Walter Lusigi, 29–38. Paris: UNESCO.

Lan, David. 1985. *Guns and rain: Guerrillas and spirit mediums in Zimbabwe.* London: James Currey.

Lance, T. 1995. "Conservation politics and resource control in Cameroon: The case of Korup National Park and its support zone." Paper presented at the African Studies Association Annual Meeting, 4 November, Orlando, Florida.

Lefebvre, Henri. 1991. *The production of space.* Translated by Donald Nicholson-Smith. Oxford: Blackwell.

Lindsey, W. K. 1987. "Integrating parks and pastoralists: Some lessons from Amboseli." In *Conservation in Africa: People, policies and practice,* edited by D. Anderson and R. Grove, 149–167. Cambridge: Cambridge University Press.

Lovett, Margot. 1994. "On power and powerlessness: Marriage and political metaphor in colonial western Tanzania." *International Journal of African Historical Studies* 27 (2): 273–301.

Low, D., and J. Lonsdale. 1976. Introduction to *History of East Africa,* edited by D. Low and A. Smith. Oxford: Oxford University Press.

Lowenthal, David, and Hugh Prince. 1964. "The English landscape." *The Geographical Review* 55 (2): 186–222.

Lowry, Alma, and T. P. Donahue. 1994. "Parks, politics, and pluralism: The demise of national parks in Togo." *Society and Natural Resources* 7:321–29.

Luanda, N. N. 1986. *European commercial farming and its impact on the Meru and Arusha peoples of Tanzania, 1920–1955.* Ph.D. diss., University of Cambridge.

Lundgren, Bjorn, and Lill Lundgren. 1972. "Comparison of some soil properties in one forest and two grassland ecosystems on Mount Meru, Tanzania." *Geografiska Annaler* 54:227–40.

Lusigi, W. 1992. "New approaches to wildlife conservation." In *Managing protected areas in Africa,* edited by Walter Lusigi, 29–38. Paris: UNESCO.

Lusigi, Walter. 1984. "Future directions for the Afrotropical Realm." In *National parks, conservation and development*, edited by J. McNeely and K. Miller, 137–146. Washington: Smithsonian Institution Press.

MacKenzie, John. 1987. "Chivalry, social Darwinism and ritualized killing: The hunting ethos in Central Africa up to 1914." In *Conservation in Africa: People, policies and practice*, edited by D. Anderson and R. Grove, 41–61. Cambridge: Cambridge University Press.

———. 1988. *The empire of nature: Hunting, conservation, and British imperialism*. Manchester: Manchester University Press.

———. 1990. Introduction to *Imperialism and the natural world*, edited by John MacKenzie, 1–14. Manchester: Manchester University Press.

Makombe, K., ed. 1993. *Sharing the land: Wildlife, people, and development in Africa*. IUCN/ROSA Environmental Series Number 1. Harrare, Zimbabwe: IUCN.

Maloba, Wunyabari. 1993. *Mau Mau and Kenya: An analysis of a peasant revolt*. Bloomington: Indiana University Press.

Marks, S. A. 1984. *The Imperial Lion: Human dimensions of wildlife management in Central Africa*. Boulder: Westview.

Mascarenhas, Adolfo. 1983. Ngorongoro: A challenge to conservation and development. *Ambio* 12 (3): 146–52.

Matzke, Gordon. 1977. *Wildlife in Tanzanian settlement policy: The case of the Selous*. Syracuse: Maxwell School of Citizenship and Public Affairs.

Mbano, B. N., R. C. Malpas, M. K. Maige, P. A. Symonds, and D. M. Thompson. 1995. "The Serengeti regional conservation strategy." In *Serengeti II: Dynamics, management, and conservation of an ecosystem*, edited by A. R. Sinclair and P. Arcese, 605–16. Chicago: University of Chicago Press.

Mbise, A. S. 1973. "The evangelist: Matayo Leveriya Kaaya." In *Modern Tanzanians: A volume of biographies*, edited by John Iliffe, 27–42. East African Publishing House.

Mbise, Ismael R. 1974. *Blood on our land*. Dar es Salaam: Tanzania Publishing House.

McNeely, J., and K. Miller, eds. 1984. *National parks, conservation, and development*. Washington: Smithsonian Institution Press.

McNeely, Jeffrey, and David Pitt, eds. 1985. *Culture and conservation: The human dimension in environmental planning*. London: Croom Helm.

Merchant, Carolyn. 1980. *The death of nature: Women, ecology, and the scientific revolution*. San Francisco: Harper and Row.

———. 1990. "Gender and environmental history." *Journal of American History* 76 (4): 1117–22.

Miller, Kenton R. 1984. "Regional planning for rural development." In *Sustaining tomorrow: A strategy for world conservation and development*, edited by F. R. Thibodeau and H. H. Field, 37–50. Hanover and London: University Press of New England.

Mitchell, Timothy. 1990. "Everyday metaphors of power." *Theory and Society* 19:545–77.

Mitchell, W. J. 1994. "Imperial Landscape." In *Landscape and power*, edited by W. J. Mitchell, 5–34. Chicago: University of Chicago Press.

Moise, Edwin. 1982. "The moral economy dispute." *Bulletin of Concerned Asian Scholars* 14 (1).

Moore, Barrington. 1966. *Social origins of dictatorship and democracy: Lord and peasant in the making of the modern world.* Boston: Beacon.

Moore, Donald. 1993. "Contesting terrain in Zimbabwe's eastern highlands: Political ecology, ethnography, and peasant resource struggles." *Economic Geography* 69:380–401.

Moore, Sally Falk, and Paul Puritt. 1977. *The Chagga and Meru of Tanzania.* London: International African Institute.

Mugera, G. M., ed. N.d. *Diseases of cattle in tropical Africa.* Nairobi: Kenya Literature Bureau.

Murphree, Marshall. 1995. *The lesson from Mahenye: Rural poverty, democracy, and wildlife conservation.* IIED Wildlife and Development Series Number 1. London: International Institute for Environment and Development.

———. 1997. "Congruent objectives, competing interests, and strategic compromise: Concept and process in the evolution of Zimbabwe's CAMP-FIRE program." Paper presented to the conference on Representing Communities: Histories and Politics of Community-Based Resource Management, June 1–3, Unicoi Lodge, Helen, Georgia.

Mustafa, K. 1993. "Eviction of pastoralists from the Mkomazi Game Reserve in Tanzania: A statement." Manuscript, International Institute for Environment and Development.

Mwijustya, V. 1985. *The dynamics of secondary vegetation in Arusha National Park, Tanzania.* Thesis, College of African Wildlife Management, Mweka, Tanzania.

Nash, Roderick. 1982. *Wilderness and the American mind.* 3rd ed. New Haven: Yale University Press.

Nelson, A. 1967. *The Freemen of Meru.* New York and Nairobi: Oxford University Press.

Neumann, Roderick P. 1992. "The Political Ecology of Wildlife Conservation in the Mount Meru Area, Northeast Tanzania." *Land Degradation and Rehabilitation* 3 (2): 85–98.

———. 1995a. "Local challenges to global agendas: Conservation, economic liberalization, and the pastoralists' rights movement in Tanzania." *Antipode* 27 (4): 363–82.

———. 1995b. "Ways of seeing Africa: Colonial recasting of African society and landscape in Serengeti National Park." *Ecumene* 2 (2): 149–69.

———. 1996. "Dukes, earls, and ersatz Edens: Aristocratic nature preservationists in colonial Africa." *Society and Space* 14:79–98.

———. 1997a. "Primitive ideas: Protected area buffer zones and the politics of land in Africa." *Development and Change* 28 (3): 559–82.

———. 1997b. "Forest rights, privileges, and prohibitions: Contextualizing state forestry policy in colonial Tanganyika." *Environment and History* 3 (1): 45–68.

Newbury, Catherine. 1994. "Introduction: Paradoxes of democratization in Africa." *African Studies Review* 37 (1): 1–8.

Ole Ntimana, W. R. 1994. "The Maasai dilemma." *Cultural Survival Quarterly* (spring): 58–59.

Omo-Fadaka, Jimoh. 1992. "The role of protected areas in the sustainable development of surrounding regions." In *Managing protected areas in Africa,* edited by Walter Lusigi, 111–16. Paris: UNESCO.

Owen, John [S.] 1962. "The National Parks of Tanganyika." In *First World Conference on National Parks,* edited by A. B. Adams, 52–59. Washington, D.C.: U.S. Government Printing Office.

———. 1963. "Awakening public opinion to the value of the Tanganyika National Parks." In *Conservation of nature and natural resources in modern African states,* edited by Gerald G. Watterson, 261–64. Morges, Switzerland: International Union for the Conservation of Nature and Natural Resources.

———. 1970. "The problem of illegal hunting in and around National Parks." Paper presented to the Judges and Magistrates Conference, 1970, Tanzania.

Peluso, N. L. 1992. *Rich forests, poor people: Forest access and control in Java.* Berkeley and Los Angeles: University of California Press.

———. 1993. "Coercing conservation? The politics of state resource control." *Global Environmental Change* 3 (2): 199–217.

Peters, P. 1995. *Dividing the commons: Politics, policy, and culture in Botswana.* Charlottesville: University of Virginia Press.

Polanyi, Karl. 1957. *The great transformation: The political and economic origins of our time.* Boston: Beacon Press.

Popkin, Samuel. 1979. *The rational peasant: The political economy of rural society in Vietnam.* Berkeley and Los Angeles: University of California Press.

Pratt, D. M., and V. H. Anderson. 1982. "Population, distribution, and behaviour of giraffe in the Arusha National Park, Tanzania." *Journal of Natural History* 16:481–89.

Prince, Hugh. 1988. "Art and agrarian change, 1710–1815." In *The iconography of landscape: Essays on the symbolic representation, design, and use of past environments,* edited by Denis E. Cosgrove and Stephen Daniels, 98–118. Cambridge: Cambridge University Press.

Pringle, Trevor R. 1988. "The privation of history: Landseer, Victoria, and the Highland myth." In *The iconography of landscape: Essays on the symbolic representation, design, and use of past environments,* edited by Denis E. Cosgrove and Stephen Daniels, 143–61. Cambridge: Cambridge University Press.

Prochaska, David. 1986. "Fire on the mountain: Resisting colonialism in Algeria." In *Banditry, rebellion, and social protest in Africa,* edited by Donald Crummey, 229–52. London: J. Currey; Portsmouth, N.H.: Heinemann.

Pudup, M. B., and M. J. Watts. 1991. "The decline of everything *or* Invasion of the culture snatchers." *Design Book Review* 22:10–13.

Pugh, Simon, ed. 1990. *Reading landscape: Country—city—capital.* Manchester: Manchester University Press.

Pullman, Robert A. 1983. "Do national parks have a future in Africa?" *Leisure Studies* 2 (1): 1–18.

Puritt, Paul. 1970. *The Meru of Tanzania: A study of their social and political organization.* Ph.D. diss., University of Illinois at Urbana-Champaign.

Ranger, Terence. 1983. "The invention of tradition in colonial Africa." In *The invention of tradition,* edited by E. Hobsbawm and T. Ranger. Cambridge: Cambridge University Press.

———. 1989. "Whose heritage? The case of Matobo National Park." *Journal of Southern African Studies* 15 (2): 217–49.

———. 1993. "The communal areas of Zimbabwe." In *Land in African agrarian systems,* edited by Thomas J. Bassett and Donald E. Crummey, 354–85. Madison: University of Wisconsin Press.

———. N.d. "Voices from the rocks." Typescript.

Redclift, Michael. 1987. "The production of nature and the reproduction of the species." *Antipode* 19 (2): 222–30.

Ricciuti, Edward. 1993. "The Elephant Wars." *Wildlife Conservation* 96 (2): 14–35.

Rodgers, W. A., R. I. Ludanga, and H. P. DeSuzo. 1977. "Biharamulo, Burigi, and Rubondo Island Game Reserves." *Tanzania Notes and Records* 81 and 82: 99–124.

Said, Edward. 1994. *Culture and imperialism.* New York: Alfred A. Knopf.

Sayer, Andrew. 1983. "Notes on geography and the relationship between people and nature." In *Society and Nature,* edited by The London Group of the Union of Socialist Geographers. London: Union of Socialist Geographers.

Sayer, Jeffrey. 1991. *Rainforest buffer zones: Guidelines for protected area managers.* Gland, Switzerland: IUCN.

Schabel, Hans. 1990. "Tanganyika forestry under German colonial administration, 1891–1919," *Forest and Conservation History* (July): 130–41.

Schroeder, R. 1993. "Shady practice: Gender and the political ecology of resource stabilization in Gambian garden/orchards." *Economic Geography* 69 (4): 349–65.

Scott, James C. 1976. *The moral economy of the peasant: Rebellion and subsistence in Southeast Asia.* New Haven: Yale University Press.

———. 1985. *Weapons of the weak: Everyday forms of peasant resistance.* New Haven: Yale University Press.

———. 1986. "Everyday forms of peasant resistance." *Journal of Peasant Studies* 13 (2): 5–35.

Shio, L. J. 1977. *A political economy of the plantation system in Arusha.* Master's thesis, University of Dar es Salaam.

Shipton, Parker, and Mitzi Goheen. 1992. "Introduction: Understanding African land-holding: power, wealth, and meaning." *Africa* 62 (3): 307–25.

Shivji, Issa. 1976. *Class struggles in Tanzania.* New York: Monthly Review Press.

———, ed. 1973. *Tourism and socialist development.* Dar es Salaam: Tanzania Publishing House.

Smith, Neil, and Phil O'Keefe. 1980. "Geography, Marx, and the concept of nature." *Antipode* 12 (2): 30–39.

Smith, Neil. 1984. *Uneven development: Nature, capital and the production of space.* New York: Basil Blackwell.

Snelson, Deborah, ed. 1987. *Arusha National Park,* Park Guidebook Series. Nairobi: Tanzania National Parks and African Wildlife Foundation.

Spear, Thomas. 1994. "Blood on the land: Stories of conquest." In *Paths toward the past,* edited by R. Harms, J. Miller, D. Newbury, and M. Wagner, 113–22. Atlanta: African Studies Association Press.

———. 1996. "Struggles for the land: The political and moral economies of land on Mount Meru." In *Custodians of the land: Ecology and culture in the history of Tanzania,* edited by G. Maddox, J. Giblin, and I. Kimambo, 213–40. London: James Currey.

SPFE. 1930a. "Minutes of the general meeting held on 4th November, 1929." *Journal of the Society for the Preservation of the Fauna of the Empire,* n.s., 10:5–6.

———. 1930b. "General meeting of the Society for the Preservation of the Fauna of the Empire." *Journal of the Society for the Preservation of the Fauna of the Empire,* n.s., 12:5–11.

———. 1933. "List of the officers of the society." *Journal of the Society for the Preservation of the Fauna of the Empire,* n.s., 18:1.

SPWFE. 1905. "Minutes of a meeting between a delegation of the SPWFE and Secretary of State for the Colonies, Lord Lyttleton." *Journal of the Society for the Preservation of the Wild Fauna of the Empire* 2:9–18.

Tanganyika National Parks. 1964. *Report and accounts of the Board of Trustees.* Arusha: East African Printers Tanganyika.

Tanganyika Territory. *Report of the Arusha-Moshi Lands Commission.* Dar es Salaam: Tanganyika Territory, 1947.

Tanzania National Parks (TANAPA). 1969. "Arusha National Park Yearly Report (Supplement) for July 1, 1968–June 30, 1969." Unpublished government document.

———. 1970. *Report and accounts of the Board of Trustees.* Arusha, Tanzania: TANAPA.

———. 1972–75. *Report and accounts of the Board of Trustees.* Arusha, Tanzania: TANAPA.

———. 1973. *Report and accounts for the years ending 30th June, 1971, 1972 and 1973.* Arusha, Tanzania: TANAPA.

———. 1994. *National policies for national parks in Tanzania.* Arusha, Tanzania: TANAPA.

Taylor, Russel. 1995. *From liability to asset: Wildlife in the Omay Communal Land of Zimbabwe.* IIED Wildlife and Development Series Number 8. London: International Institute for Environment and Development.

Tewa, T. S. 1963. "The value of the tourist industry in the conservation of natural resources in Tanganyika." In *Conservation of nature and natural resources in modern African states,* edited by Gerald G. Watterson, 336–39. Morges, Switzerland: International Union for the Conservation of Nature and Natural Resources.

Thomas, Stephen. 1995. *Share and share alike? Equity in CAMPFIRE.* IIED Wildlife and Development Series Number 2. London: International Institute for Environment and Development.

Thompson, E. P. 1967. "Time, work-discipline, and industrial capitalism." *Past and Present* 38:56–97.

——— . 1971. "The moral economy of the English crowd in the eighteenth century." *Past and Present* 50:76–136.

——— . 1975. *Whigs and hunters: The origin of the Black Act*. New York: Pantheon Books.

——— . 1991. "The moral economy reviewed." In *Customs in common*, 259–351. London: Penguin Books.

Thorsell, J. 1982. "Evaluating effective management in protected areas: An application to Arusha National Park, Tanzania." Paper presented at the World National Parks Congress, Bali, Indonesia, 18–20 October 1982.

Torgovnick, Marianna. 1990. *Gone primitive: Savage intellects, modern lives*. Chicago: University of Chicago Press.

Turshen, Meredeth. 1984. *The political ecology of disease in Tanzania*. New Brunswick, N.J.: Rutgers University Press.

Turton, David. 1987. "The Mursi and national park development in the Lower Omo Valley." In *Conservation in Africa: People, policies and practice*, edited by David Anderson and Richard Grove, 169–86. Cambridge: Cambridge University Press.

United Republic of Tanzania. 1964. *Annual report of the Ngorongoro Conservation Unit*. Dar es Salaam: Ministry of Agriculture, Forests and Wildlife.

——— . 1965. *Annual report of the Ngorongoro Conservation Unit*. Dar es Salaam: Ministry of Agriculture, Forests and Wildlife.

——— . 1966. *Annual report of the Ngorongoro Conservation Unit*. Dar es Salaam: Ministry of Agriculture, Forests and Wildlife.

——— . 1967. *Annual report of the Ngorongoro Conservation Unit*. Dar es Salaam: Ministry of Agriculture, Forests and Wildlife.

——— . 1981. *1978 population census*. Dar es Salaam: Bureau of Statistics.

——— . 1988. *Hotels and national parks, 1987*. Dar es Salaam: Bureau of Statistics.

——— . 1989. *1988 population census*. Dar es Salaam: Bureau of Statistics.

——— . 1992. *Report of the Presidential Committee of Inquiry into land matters*. Vols. 1 and 2. Dar es Salaam: United Republic of Tanzania, Government Printers.

Vesey-FitzGerald, D. F. 1967. "A glimpse of Eden." *Africana* 3 (3): 11–15.

——— . 1973. "The dynamic aspects of the secondary vegetation in Arusha National Park, Tanzania." *East African Agricultural and Forestry Journal* (January): 314–27.

——— . 1974. "Utilization of the grazing resources by buffaloes in the Arusha National Park, Tanzania." *East Africa Wildlife Journal* 12: 107–34.

Warren, Louis. 1994. *Poachers and conservationists: State power, local resistance, and the history of the American West*. Paper presented at the Agrarian Studies Colloquium, Yale University, September 16.

Watterson, Gerald G. 1963. "Origin and aims." In *Conservation of nature and natural resources in modern African states*, edited by Gerald G. Watterson, 9–11. IUCN Publications New Series 1. Morges, Switzerland: International Union for the Conservation of Nature and Natural Resources.

Watts, Michael. 1983. *Silent violence: Food, famine, and peasantry in northern Nigeria*. Berkeley and Los Angeles: University of California Press.

―――. 1988. "Struggles over land, struggles over meaning." In *A ground for a common search,* edited by Reginald Golledge, 31–51. Santa Barbara: University of California at Santa Barbara Press.

Weiskel, Timothy. 1987. "Agents of empire: Steps toward an ecology imperialism." *Environmental Review* 11 (4): 275–88.

Wells, M., and K. Brandon. 1992. *People and parks: Linking protected area management with local communities.* Washington, D.C.: World Bank, WWF, USAID.

―――. 1993. "The principles and practice of buffer zones and local participation in biodiversity conservation." *Ambio* 22 (2–3): 157–62.

Wells, Roger. 1994. "E. P. Thompson, *Customs in common,* and Moral Economy." *Journal of Peasant Studies* 21 (2): 263–307.

Western, David. 1982. "Amboseli National Park: Enlisting landowners to conserve migratory wildlife." *Ambio* 11:302–08.

Williams, Dale. 1984. "Morals, markets, and the English crowd in 1766." *Past and Present* 104:56–73.

Williams, Raymond. 1973. *The country and the city*. London: Hogarth Press.

―――. 1980. *Problems in materialism and culture: Selected essays*. London: NLB.

Wilson, Ken. 1997. "Of diffusion and context: The bubbling up of community-based resource management in Mozambique." Paper presented to the conference on Representing Communities: Histories and Politics of Community-Based Resource Management. June 1–3, Unicoi Lodge, Helen, Georgia.

Wolf, Eric R. 1966. *Peasants*. Englewood Cliffs, N.J.: Prentice-Hall.

―――. 1982. *Europe and the people without history*. Berkeley and Los Angeles: University of California Press.

Worster, Donald. 1985. *Nature's economy: A history of ecological ideas*. Cambridge: Cambridge University Press.

―――. 1990. "Transformations of the earth." *Journal of American History* 76 (4): 1087–106.

WWF. 1990. *Proposed WWF Tanzania Country Programme*. Typescript. Washington, D.C.: World Wide Fund for Nature.

Yeager, Rodger, and Norman Miller. 1986. *Wildlife, wild death: Land use and survival in Eastern Africa*. Albany: State University of New York Press.

Index

Abraham (a headman), 71, 72, 73
Achoo (a Meru clan), 54
Adam, Y., 42
Adams, Jonathan S., 48–49
ADMADE (Administrative Management Design for Game Management Areas, Zambia), 207
Africa: agrarian crises in, 39; Edenic image of, 3–4, 17–18, 34, 128–29, 177; increased tensions in, 5–6; landscape vision of, 1; militarization in, 6; politics of national parks in, 31–32, 33, 34–37 (*see also specific parks*); protected areas in, 4
African Freedom Army, 135
African Land Army, 135
African liberation movements, 31–32. *See also* nationalist movement
Africans. *See specific peoples*
African Special Project, 139–40
African Wildlife Foundation (AWF), 209, 227–28n26
African Wildlife Leadership Foundation (AWLF), 142, 150
agriculture, and soil conservation, 43
Akheri (Tanzania), 60, 64, 77, 78, 96
Akram-Lodhi, A. Haroon, 44, 45
Algeria, 47
American romanticism, 16, 24
Ames, Evelyn, 2, 177

A-MLC (Arusha-Moshi Lands Commission), 69–70
Anderson, David, 18, 34, 40–41, 47
Annals of the Association of American Geographers, 27
antelope hunting, 163–64
Arusha Manifesto, 140
Arusha-Moshi Lands Commission (A-MLC), 69–70
Arusha National Park (Tanzania): administration's responsiveness to villagers, 191–92, 196–97; beekeeping in, 153, 156, 187–88; boundaries of, 2, 12 (map), 92, 149–50, 150 (map), 165–67; and CCS, 209–11; creation of, 148–56, 150–51 (maps); customary rights eliminated at, 151–56; ecology of, 158–60, 161, 168–70; economic benefits to Africans, 167, 210; Edenic/scenic view of, 2, 177, 202; European vs. Meru view of, 50, 177–78; guards, abuses by, 189–90, 197–98, 229n26; guards, salaries of, 198–99, 230n49; land alienation at, 2, 76, 150–51, 151 (map); management/planning of, 160–62, 226–27n5; natural-resource crimes at, 162–67; policies criticized as colonialist, 76, 194; and relations of mutual obligation/reciprocity, 44, 195, 196–97, 199–201; right-of-way

tlement patterns of, 216n53; timber rights of, 115–16, 219n70. *See also* Mount Meru

Ministry of Agriculture, Forests and Wildlife (Tanzania), 143

Ministry of Information and Tourism (Tanzania), 143

Ministry of Natural Resources and Tourism, 147

missionaries: Lutheran, 55–56, 60–61, 63–64, 65 (fig.), 112; schools, 64, 65 (fig.); view of Africans as natural Christians, 141

Mitchell, Sir Philip, 104, 115

Mitchell, Timothy, 41, 45–46, 47–48

Mitchell, W. J., 17, 32

Miwok people (Yosemite Valley), 30

mob justice, 48–49

Molel (Meru villager), 82–83, 84, 85–86, 87

Momella Farm (Trappe farm; Arusha National Park, Tanzania): inclusion in Arusha National Park, 148, 149, 150, 166–67; Nasula claims to, 182–83; proposed allocation of for Africans, 215n21, 215n27; squatters on, 82–83; as underutilized, 67

Momella Game Lodge (Arusha National Park, Tanzania), 167

Momella Lakes (Mount Meru, Tanzania), 149, 158, 159, 226n105

Monas, A. S., 82, 83, 84, 174, 175

Montagne d'Ambre National Park (Madagascar), 5

Moore, Barrington, 39

Moore, Donald, 50, 228–29n2

moral community, 195–96

moral economy, landed, 11, 37–44, 51–96, 174–205; as anti-capitalist, 42; and conservation, 109–11; and customary rights/land claims, 175–76, 180–81, 187–88; and encroachment/tenure security, 13, 182–85; and father/son tensions, 96; and interpretation of tradition, 43–44; of *kihamba* land, 80–81, 96; and landowner/peasant relationship, 178–79; and local officials' collaboration with villagers, 185–87; vs. market economy, 41, 214n5; park policies as violating, 13, 187–95; and patronage, 40, 41, 196; of peasant societies, 38–40, 214n4; and relations between

park staff and villagers, 195–205, 198 (fig.); and self-interest, 41; and subsistence ethic vs. risk-bearing, 39–40, 41–42, 43, 44, 96, 214n4; E. P. Thompson on, 37–38, 41, 42. *See also* Meru Land Case; Meru people; Nasula Village; Ngongongare Kitongoji

Moran, Thomas, 24

Mount Kilimanjaro (Tanzania), 90, 123, 124, 144, 225n97

Mount Meru Complete Game Reserve, 112, 124

Mount Meru (Tanzania), 51–96; ancient human occupation of, 159; Boers' use of wildlife on, 61–62; cultural meaning/identity for the Meru, 50, 54–55, 56, 153–54, 154 (fig.), 177–78, 211; designation/potential as a national park, 123, 149; ecology/geology of, 158–59; evolution of state conservation policy on, 111–21; game and forest reserves, 61, 63, 64 (map), 68 (table) (*see also* conservation); German colonization of, 54, 60–63; labor-exporting/importing regions, 65–66; land alienation on, 2, 11, 61–63, 64 (map), 66; land/resource control on, 50; as picturesque, 2, 3 (fig.); precolonial land/production on, 56–60, 214n9; precolonial Meru migrations to, 53–56; rainfall/climate, 58, 63, 158; as refuge, 54; timber rights on, 111–12, 115–16, 219n70; travel writing about, 177. *See also* Arusha National Park; Meru people

Mwijustya, V., 159

Mwinyi, Ali Hassan, 77, 144

myths: Edenic, 3–4, 17–18, 32–33, 34, 128–29, 177, 202; of emptiness, 29, 214n2; European vs. African, 177–78, 229n2; of primitive Africans, 18, 128, 134, 222n32

Nairobi, 135

Nako (a Meru clan), 55

Namibia, 4

Nanyaro (a Meru clan), 53

Nasary (a Meru clan), 54

Nash, Roderick, 33–34

Nasula Village (Olkangwado Kijiji, Tanzania), 11, 81–90; cultivation pattern in, 88–89, 89 (fig.); drinking water in, 90; ecology of, 159; farming in, 89–90;

Text: 10/13 Galliard
Display: Galliard